FIRST EDITION

Problems Book for Organic Chemistry

Robert Engel, A. David Baker, and JaimeLee Iolani Rizzo

Queens College of CUNY

Bassim Hamadeh, CEO and Publisher
John Remington, Executive Editor
Gem Rabanera, Project Editor
Chelsey Schmid, Production Editor
Jess Estrella, Senior Graphic Designer
Trey Soto, Licensing Coordinator
Natalie Piccotti, Director of Marketing
Kassie Graves, Vice President of Editorial
Jamie Giganti, Director of Academic Publishing

Printed in the United States of America.

ISBN: 978-1-5165-2834-9 (pbk) / 978-1-5165-2835-6 (br)

CONTENTS

Introduction 1

ORGANIC CHEMISTRY I **3**

First Examination (I) 3

First Examination (II) 4

First Examination (III) 5

Second Examination (I) 7

Second Examination (II) 9

Second Examination (III) 11

Third Examination (I) 13

Third Examination (II) 15

Third Examination (III) 17

Final Examination (I) 19

Final Examination (II) 22

Final Examination (III) 24

ORGANIC CHEMISTRY II **27**

First Examination (I) 27

First Examination (II) 29

First Examination (III) 31

Second Examination (I) 35

Second Examination (II) 37

Second Examination (III) 39

Third Examination (I) 43

Third Examination (II) 45

Third Examination (III) 47

Final Examination (I) 49

Final Examination (II) 52

Final Examination (III) 55

ANSWERS **59**

Organic Chemistry I ***61***

First Examination (I) 61

First Examination (II) 65

First Examination (III) 69

Second Examination (I) 73

Second Examination (II) 76

Second Examination (III) 79

Third Examination (I) 83

Third Examination (II) 85

Third Examination (III) 88

Final Examination (I) 91

Final Examination (II) 97

Final Examination (III) 101

Organic Chemistry II ***105***

First Examination (I) 105

First Examination (II) 108

First Examination (III) 111

Second Examination (I) 115

Second Examination (II) 118

Second Examination (III) 121

Third Examination (I) 125

Third Examination (II) 128

Third Examination (III) 130

Final Examination (I) 133

Final Examination (II) 137

Final Examination (III) 141

INTRODUCTION

The sets of problems and solutions contained herein are intended to be a supplement to a regular textbook used in a standard two-semester organic chemistry course as is taught for chemistry, chemical engineering, and pre-health professions students. It is intended as a review and practice aide directing the students toward the solving of problems typical of such courses of study and the real-life aspects of the indicated disciplines.

The format of this book is as a series of timed examinations covering the topics as normally are taught in the normal two-semester sequence of organic chemistry. The student should perform the solving of the problems in the manner used in performing examinations, a timed format for completion of the problems with fully annotated answers. The students can subsequently "grade" their performances on these examinations to assess their abilities in mastering the topic. The "examinations" are provided in sequence as one would progress through the two-semester sequence, with "final" (comprehensive) examinations at the end of each semester's effort. Best results are obtained if the provided answers are reviewed only after completing each examination. Several sets of examinations are given for each stage of progress through the program.

Read each question before you attempt to answer it. The value of each question is noted, with a total of 100 points for the entire examination. You may use a calculator, if you find it necessary. Time allowed = 100 minutes.

1. (10 points) Draw a complete structure showing all atoms and bonds for each of the compounds named below.
   (a) 2-chloro-3,4-dimethylhexane
   (b) 1-bromo-3-ethylpentane

2. (20 points) Ammonia (NH_3) reacts with boron trifluoride (BF_3) to form an adduct (H_3NBF_3). Using molecular orbital concepts, show the orbitals of each reacting species involved in this addition and describe the resultant bond in the adduct. You should draw an energy diagram indicating the conversion of reactant orbitals to product orbitals, and give a brief explanation of the reaction process and the new bond that is formed.

3. (20 points) Calculate the weights of CO_2 and H_2O that would be formed by complete combustion of a sample of benzyl alcohol (formula C_7H_8O) that weighs 6.00 mg. (Atomic masses: C, 12.011; H, 1.008; O, 15.999)

4. (20 points) You have a mixture of two solids, sodium chloride (ordinary table salt, NaCl) and triacontane (the straight-chain alkane of formula $C_{30}H_{62}$). Describe how you would use water to separate these two solids from each other and the principle upon which your separation procedure is based.

5. (15 points) Consider the molecule 1,2-dichloroethane. Using Newman projections, represent the two staggered conformations of this molecule. Which of these two conformations would you expect to be present in greater amount in a gaseous sample of 1,2-dichloroethane at 100°C? Explain your decision.

6. (15 points) Arrange each of the following sets of compounds in order of increasing boiling temperature.
   (a) ethanol, ethane, 1-butanol
   (b) 1-pentanol, diethyl ether, pentane
   (c) 3,3-diethylpentane, 1-bromononane, nonane

Read each question before you attempt to answer it. The value of each question is noted, with a total of 100 points for the entire examination. You may use a calculator, if you find it necessary. Time allowed = 100 minutes.

1. (20 points) A compound known to contain only carbon and hydrogen with no π bonds is found by combustion analysis to be composed of 87.19% carbon and 12.81% hydrogen. Other experiments show it to have an approximate molecular weight of 110. Give the molecular formula for the compound, and tell how many rings are present within its structure.

2. (10 points) For the named compounds (a) and (b) give complete structures showing all of the atoms and bonds.
   (a) 3,4-dibromo-5-ethyl-2-methyloctane
   (b) 3-ethyl-2-hexanol

3. (18 points) Using an sp^3 orbital on carbon and a p orbital on bromine, construct the molecular orbital energy diagram and show the shapes of the resultant molecular orbitals for the C-Br bond in $H_3C\text{-}Br$.

4. (20 points) Consider the molecule $BrCH_2\text{-}CH_2Cl$. Draw Newman projections of this molecule as viewed along the carbon-carbon bond which illustrate each of the following:
   (a) the lowest energy conformation for the molecule.
   (b) the highest energy conformation for the molecule.
   (c) an eclipsed conformation intermediate in energy between the highest and lowest energy conformations.
   (d) a staggered conformation intermediate in energy between the highest and lowest energy conformations.

5. (15 points) Give the hybridization for each of the carbon atoms indicated (a)-(e) in the molecule shown below.

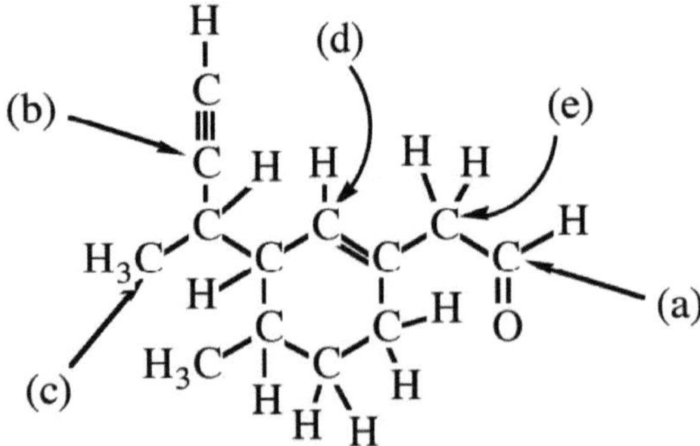

6. (17 points) Upon chlorination (with Cl_2 and light) the molecule 2-methylpentane forms five different monochlorinated products (formula $C_6H_{13}Cl$ for each). Give the structure and the IUPAC name for each monochlorinated product.

ORGANIC CHEMISTRY I FIRST EXAMINATION (III)

Read each question before you attempt to answer it. The value of each question is noted, with a total of 100 points for the entire examination. You may use a calculator, if you find it necessary. Time allowed = 100 minutes.

1. (20 points) In a concise, but complete, manner explain why high energy input is required to cause rotation about the C3-C4 bond of *trans*-3-hexene, but interconversion of the conformations of hexane (rotation about the C3-C4 bond) occurs readily even at low temperatures.

2. (20 points) Using an sp^2 and a p orbital from each of the two carbon atoms, construct the molecular orbital diagram and show the shapes of the resultant molecular orbitals for the σ and π carbon-carbon bonds of ethene ($H_2C{=}CH_2$).

3. (15 points) Give a brief explanation why the Newman projection shown below for HO-CH_2-CH_2-OH *does not* represent the lowest energy conformation for the molecule.

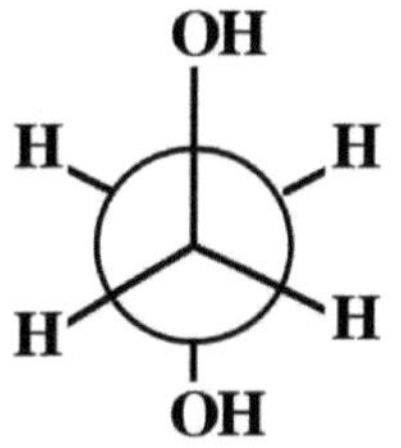

4. (20 points) Draw the structures of *all* of the compounds which can have the formula C_4H_7Cl. (There are 16 structures to be drawn.)

5. (15 points) For each of the following pairs of isomers, tell which would liberate the least amount of energy upon complete combustion. Explain your decisions.

   (a) 3,3-dimethylpentane or heptane
   (b) *cis*-1,2-dimethylcyclopentane or *trans*-1,2-dimethylcyclopentane
   (c) cyclohexane or propylcyclopropane

6. (10 points) Using combustion analysis for carbon and hydrogen, would you be able to distinguish between samples of cyclohexane and cycloöctane? Explain why or why not.

ORGANIC CHEMISTRY I SECOND EXAMINATION (I)

Read each question before you attempt to answer it. The value of each question is noted, with a total of 100 points for the entire examination. You may use a calculator or molecular models if you find them necessary. Time allowed = 100 minutes.

1. (20 points) Give the structures of the major organic product(s) in each of the following reactions. If no reaction occurs, so indicate.

(a) [cyclohexyl–CH(OH)–] OH $\xrightarrow[H_2SO_4]{CrO_3}$

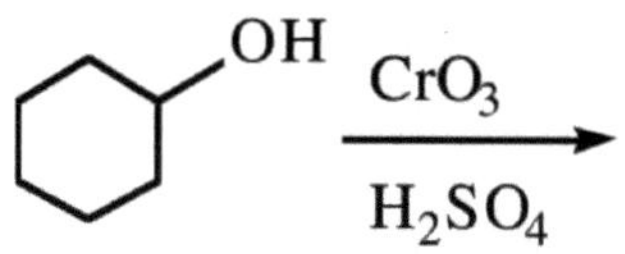

(b) [cyclopentyl–CH_2CH_2–OH] $\xrightarrow[H_2SO_4]{CrO_3}$

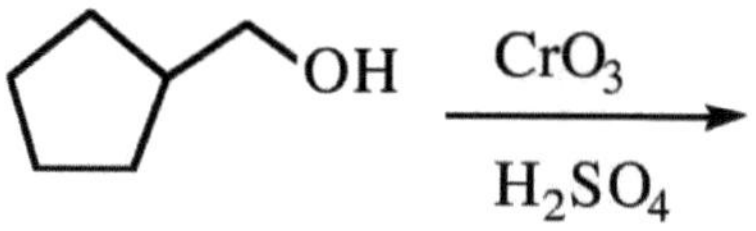

(c) $(CH_3)_3C\text{-}OH \xrightarrow{H_2SO_4}$

(d) $(CH_3)_2CH\text{-}CH(Br)\text{-}CH_3 \xrightarrow{K^{+}\,{}^{-}OC(CH_3)_3}$

(e) $(CH_3)_3C\text{-}OH \xrightarrow[KOH,\ H_2O]{KMnO_4}$

2. (20 points) Write out the *complete* mechanism using the curved arrow formalism for the formation of CH_3OCH_3 upon treatment of methanol (CH_3OH) with sulfuric acid (H_2SO_4).

3. (18 points) For each of the following, complete the structure on the right so that it represents the same stereoisomers as represented by the structure on the left.

(a)

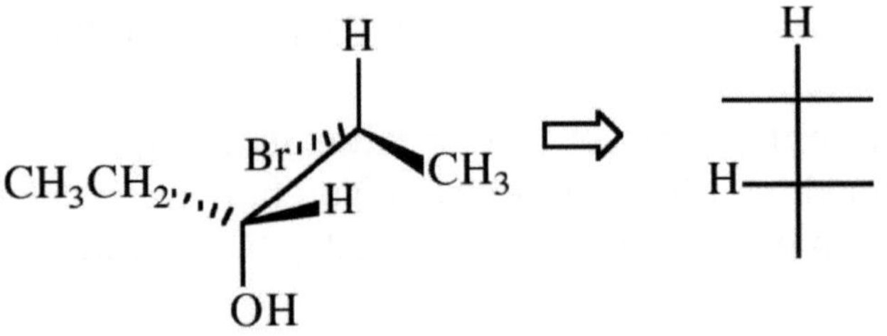

H
CH_3CH_2 Br H CH_3 H
OH

H
H—

(b)

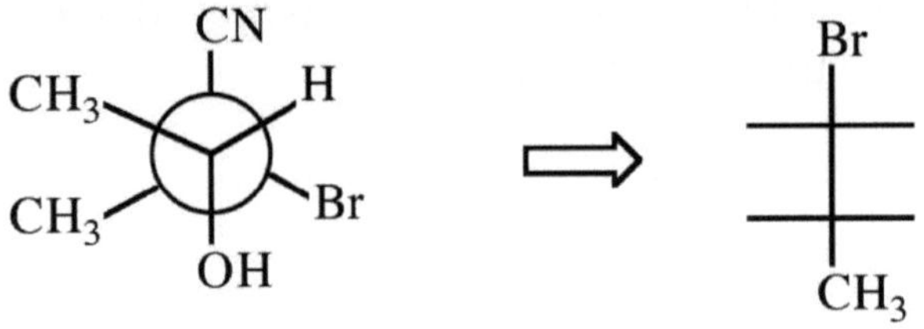

CN
CH_3 H
CH_3 Br
OH

Br
CH_3

(c)

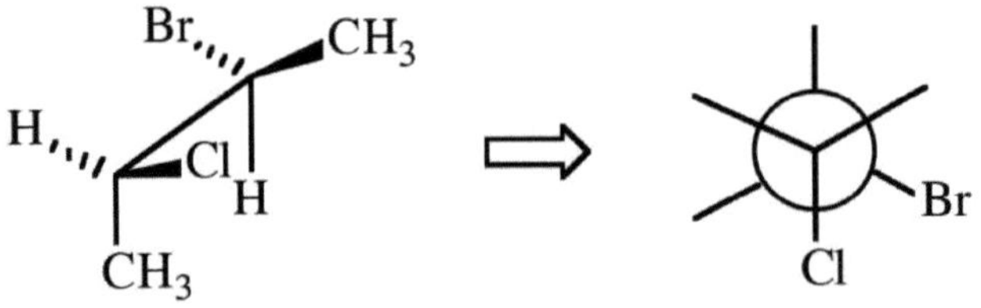

Br CH_3
H Cl H
CH_3

Br
Cl

4. (16 points) Draw a "wedge" structure representing correctly the stereochemistry for each of the following:

(a) (*R*)-3-iodohexane
(b) (*S*)-1-bromo-2-butanol

5. (14 points) Give the required reagents for each of the following conversions:

(a) $(CH_3)_3CH \longrightarrow (CH_3)_3C\text{-}Br$

(b) $CH_3CH_2CH_2CH_2OH \longrightarrow CH_3CH_2CH_2\text{-}\overset{\displaystyle O}{\overset{\|}{C}}\text{-}H$

6. (12 points) Give a brief explanation as to why the conversion shown below would *not* give a good yield of the intended product.

$$(CH_3)_2CH\text{-}\underset{}{\overset{\displaystyle OH}{\overset{|}{C}}}H\text{-}CH_3 \xrightarrow[H_2O]{HBr} (CH_3)_2CH\text{-}\overset{\displaystyle Br}{\overset{|}{C}}H\text{-}CH_3$$

ORGANIC CHEMISTRY I SECOND EXAMINATION (II)

Read each question before you attempt to answer it. The value of each question is noted, with a total of 100 points for the entire examination. You may use a calculator or molecular models if you find them necessary. Time allowed = 100 minutes.

1. (20 points) For each of the following, complete the unfinished structures so they represent the same stereoisomers as depicted on the left.

(a) H_3C OH H H_3C H Br ⇨ OH Br

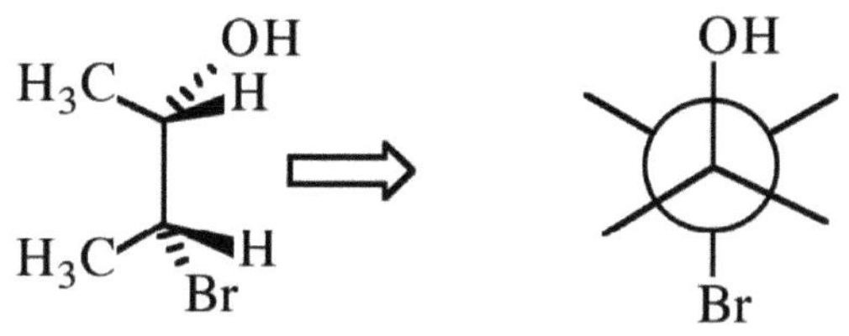

(b) H_3C H Br CH_2CH_3 ⇨ H ⇨ Br CH_3

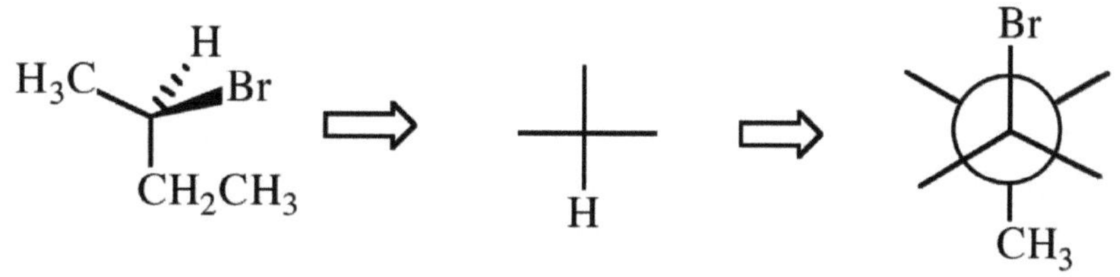

(c) CH_3 Cl CH_2CH_3 H Br Cl ⇨ Br CH_2CH_3

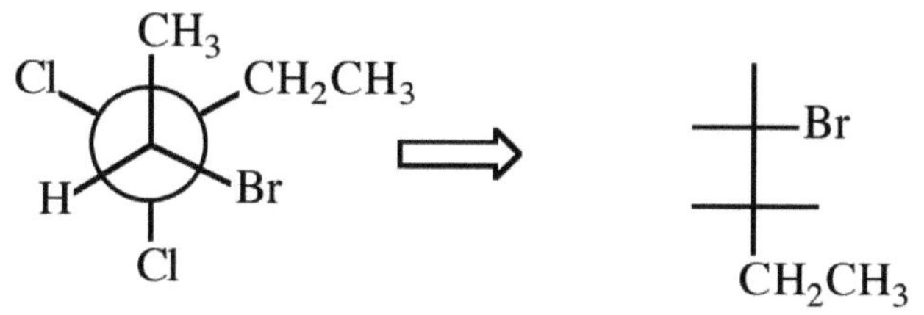

2. (18 points) Concisely define each of the following terms:
(a) activated complex
(b) *meso*-compound
(c) diastereoisomers

3. (20 points) Give a concise explanation utilizing molecular orbital concepts for the observed greater stability of the *t*-butyl cation (**A**) as compared to the 1-butyl cation (**B**).

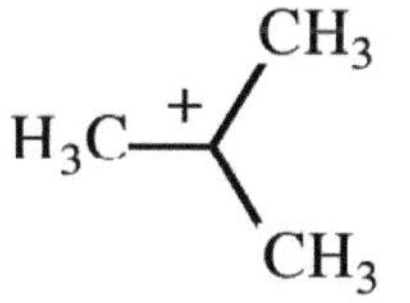

CH_3 H_3C + CH_3 **A**

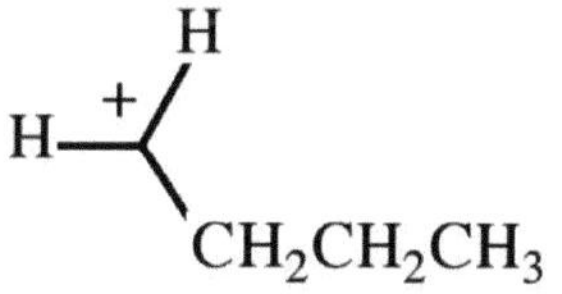

H H + $CH_2CH_2CH_3$ **B**

4. (15 points) Provide the reagents and reaction conditions required for each of the conversions shown below.

(a)

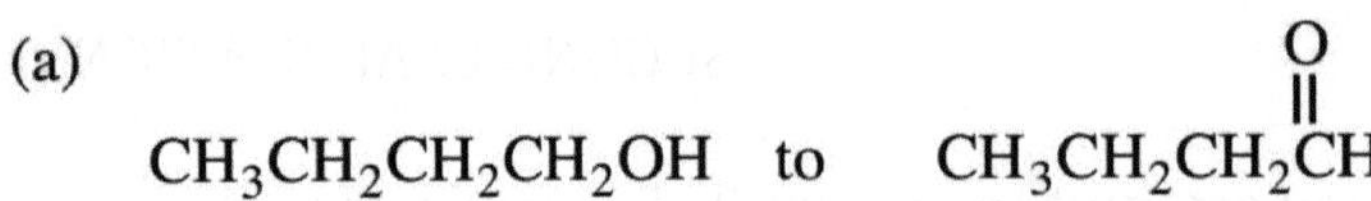

(b)

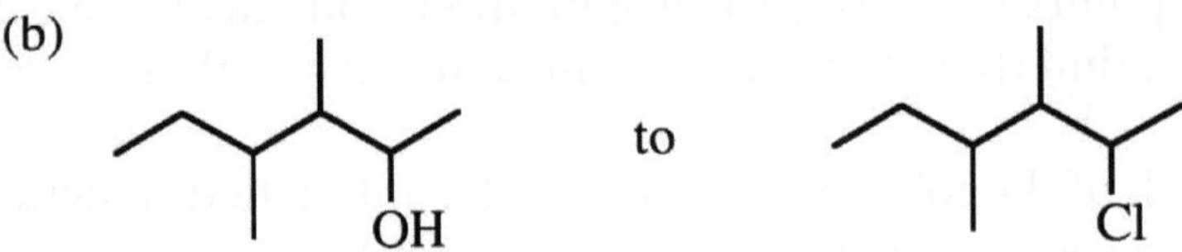

(c)

Br

to

5. (15 points) Using the curved arrow formalism and showing any and all true intermediates, give the complete mechanism for the formation of ICH_3 upon treating $HOCH_3$ with sulfuric acid and sodium iodide.

6. (12 points) Give the structure for each of the compounds indicated **C-F**.

C ($C_7H_{16}O$) $\xrightarrow{PBr_3}$ **D** ($C_7H_{15}Br$) $\xrightarrow{K^{+\,-}OC(CH_3)_3}$ **E** (C_7H_{14})

C $\xrightarrow{KMnO_4,\ KOH,\ H_2O}$ (ketone, O)

$\xrightarrow[h\nu]{Br_2}$ **F** ($C_7H_{15}Br$) $\xrightarrow{K^{+\,-}O(CH_3)_3}$ **E**

ORGANIC CHEMISTRY I — SECOND EXAMINATION (III)

Read each question before you attempt to answer it. The value of each question is noted, with a total of 100 points for the entire examination. You may use a calculator or molecular models if you find them necessary. Time allowed = 100 minutes.

1. (15 points) For each of the following, complete the unfinished structures so they represent the same stereoisomers as depicted on the left.

(a)

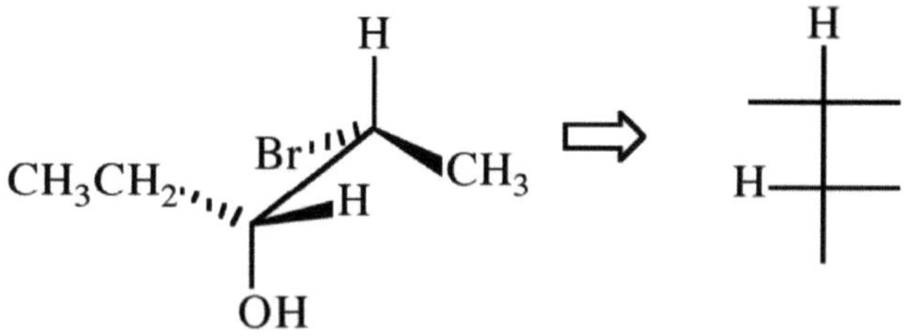

(b)

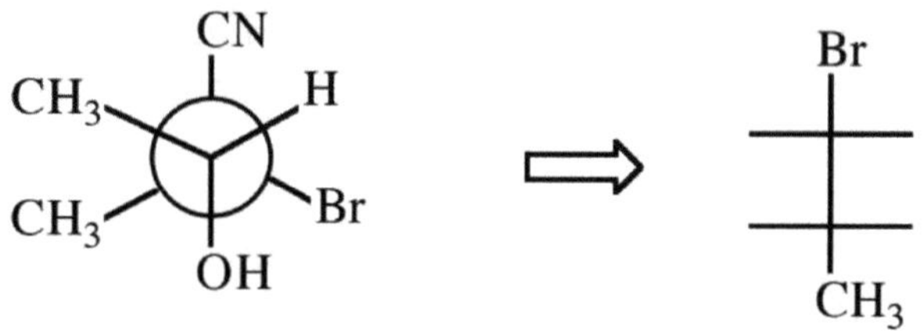

(c)

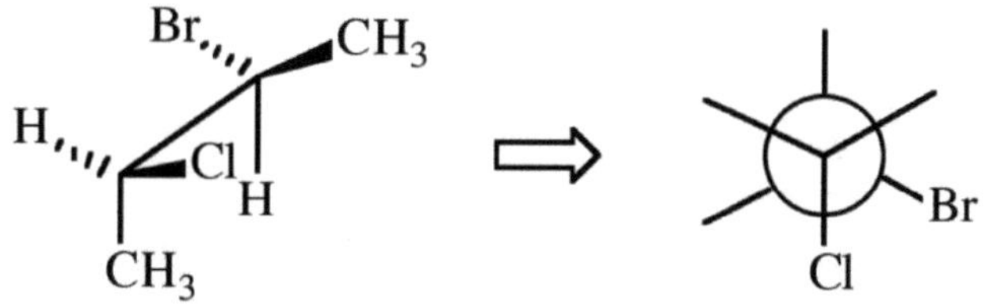

2. (15 points) Concisely define each of the following terms:
(a) E2 Reaction
(b) *syn*-hydroxylation
(c) nucleophile

3. (21 points) Utilizing **molecular orbital** concepts, explain why the compound shown below (1-bromobicyclo[2.2.1]heptane) fails to undergo facile substituion *or* elimination reactions involving the bromine at position-1.

4. (12 points) Concisely state what is wrong with each of the attempted syntheses shown below. You should tell *why* the indicated product would not be isolated in good yield and what actually *would* happen.

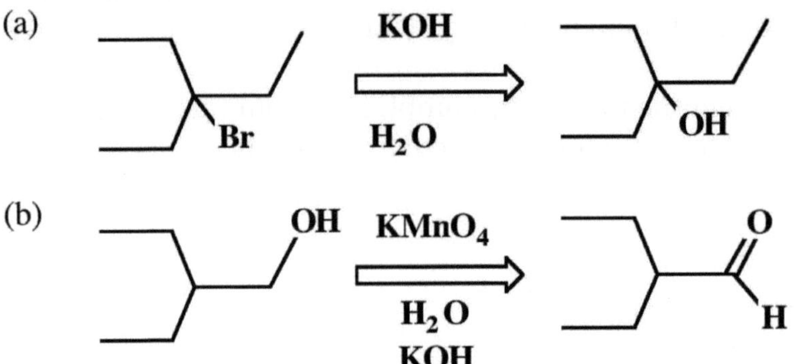

5. (12 points) Draw a reaction progress diagram for the conversion shown below. You should indicate the structure of any intermediates, the relative energies of reactants, products, and any intermediates (should there be any), and the significance of energy differences between such species.

$$(CH_3)_3COH + HBr \longrightarrow (CH_3)_3CBr + H_2O$$

6. (25 points) Give the structure for each of the compounds indicated **A-E**.

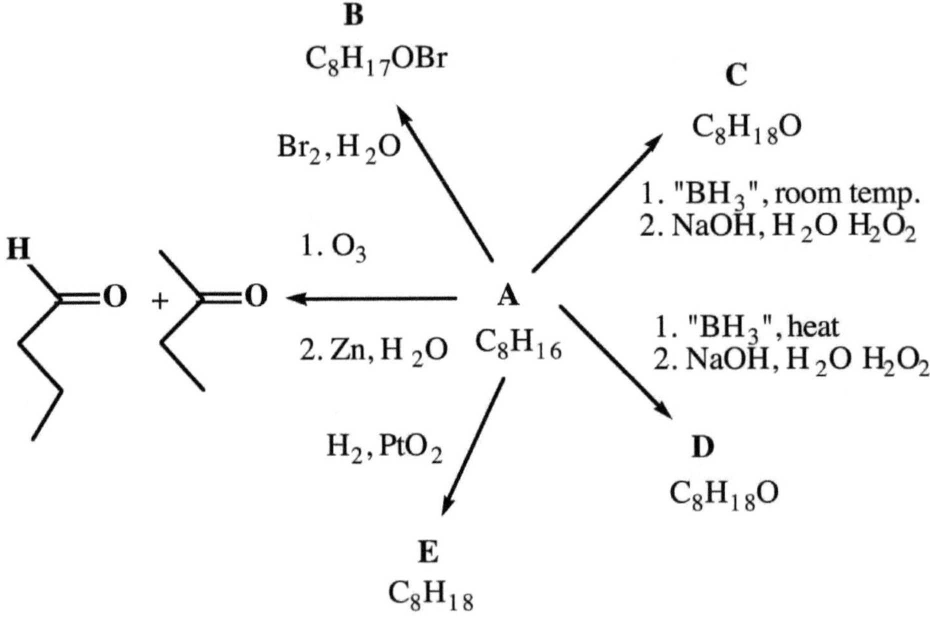

ORGANIC CHEMISTRY I THIRD EXAMINATION (I)

Read each question before you attempt to answer it. The value of each question is noted, with a total of 100 points for the entire examination. You may use a calculator or molecular models if you find them necessary. Time allowed = 100 minutes.

1. (25 points) Give the reagents required for the conversion of 1-methylcyclopentene into each of the compounds listed **A-E**. Caution: more than one step *may* be required for each conversion.

CH_3 D **A** CH_3 OH **B** CH_3 H OH HO **C** CH_3 HO OH H **D** CH_3 **E**

2. (12 points) Draw the two chair forms of 1,1,3-trimethylcyclohexane and tell which is the more stable and which is the less stable form.

3. (25 points) Give the structure for each of the compounds indicated **F-J**.

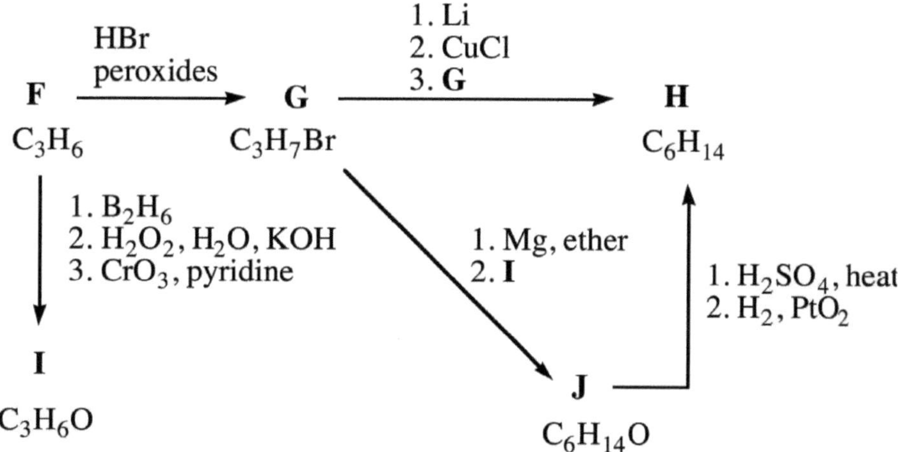

4. (25 points) Predict the major organic product in each of the following reactions.

(a) (*R*)-3-hexanol $\xrightarrow{SOCl_2\text{, dioxane}}$

(b) 3-ethyl-3-bromopentane $\xrightarrow{KOC(CH_3)_3}$

(c) 3-ethyl-3-bromopentane $\xrightarrow{H_2O\text{, ethanol}}$

(d) 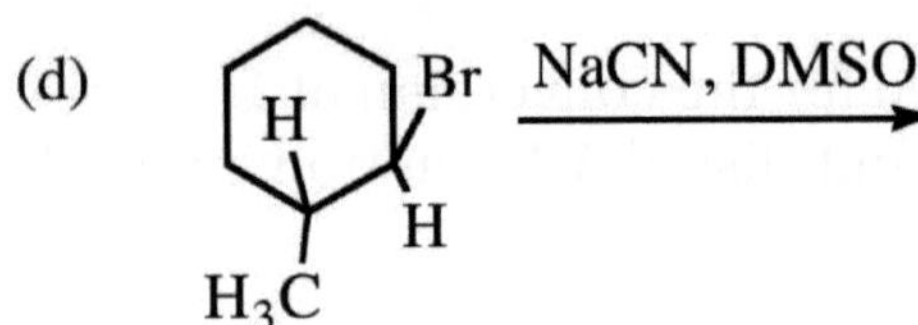

(e) 1-bromohexane $\xrightarrow{NaOCH_2CH_3}$

5. (13 points) The overall reaction

$$2\,A + B \rightarrow D + E$$

is found to observe the rate law

$$\text{Rate} = k[A][B].$$

Write a mechanism for the reaction that is consistent with the observed rate law.

Read each question before you attempt to answer it. The value of each question is noted, with a total of 100 points for the entire examination. You may use a calculator or molecular models if you find them necessary. Time allowed = 100 minutes.

1. (25 points) Give the reagents required for the conversion of (*Z*)-3-methyl-3-hexene into each of the compounds listed **A-D**. More than one step may be required for each conversion.

A OH

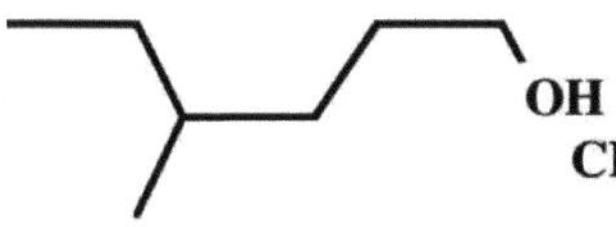

B OH OH CH_3CH_2 H CH_3 CH_2CH_3 + enantiomer (racemic)

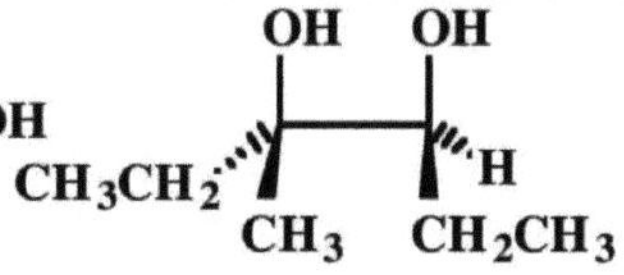

C OH Br CH_3CH_2 H CH_3 CH_2CH_3 + enantiomer (racemic)

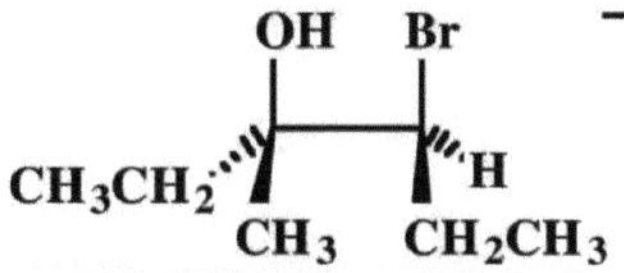

D OH racemic

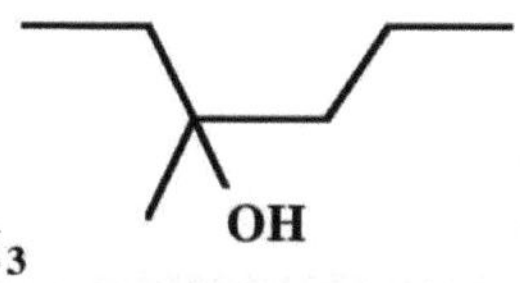

2. (20 points) Give a concise explanation (words and structures) why no β-elimination reaction occurs with the haloalkane shown below when it is treated with potassium *t*-butoxide.

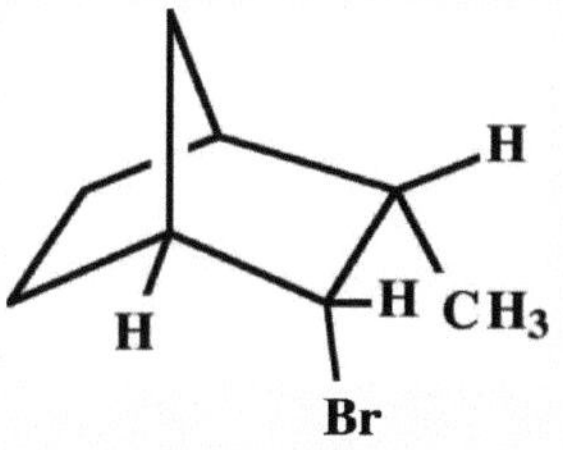

H H CH_3 H Br

3. (30 points) Give the structure for each of the compounds indicated **E-J**.

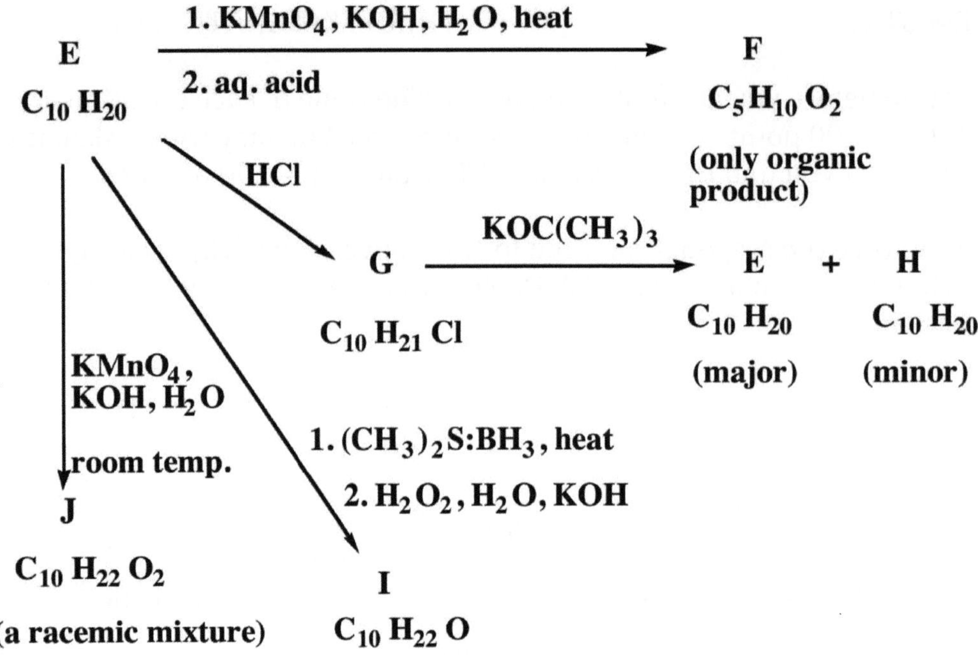

4. (25 points) Consider the reaction of (*R*)-2-hexanol with $SOCl_2$ in dioxane solvent (shown below). Give an explanation for the predominant formation of (*R*)-2-chlorohexane in this reaction.

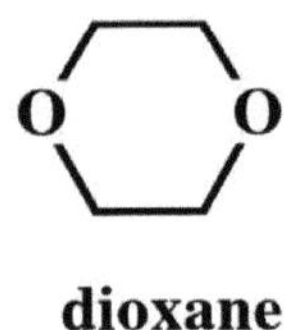

dioxane

ORGANIC CHEMISTRY I THIRD EXAMINATION (III)

Read each question before you attempt to answer it. The value of each question is noted, with a total of 100 points for the entire examination. You may use a calculator or molecular models if you find them necessary. Time allowed = 100 minutes.

1. (25 points) Give the reagents required for the conversion of 1-methylcyclohexene into each of the compounds listed **A**-**D**. More than one step may be required for each conversion.

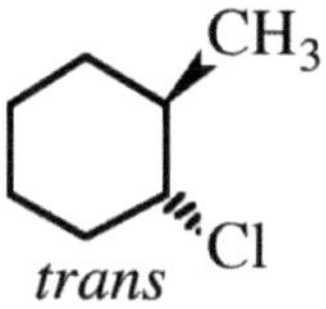

CH_3
Cl

trans

A

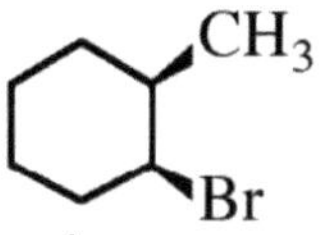

CH_3
Br

cis

B

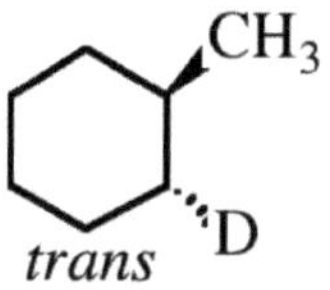

CH_3
D

trans

C

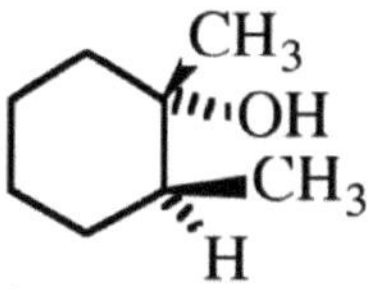

CH_3
OH
CH_3
H

D

2. (25 points) Give the structure of the major organic product in each of the following reactions. Be careful to show the proper stereochemistry when appropriate.

(a) 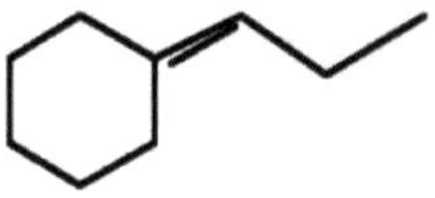 1. B_2H_6, heat; 2. H_2O_2, KOH, H_2O →

(b) (*R*)-2-iodohexane excess NH_3 →

(c) 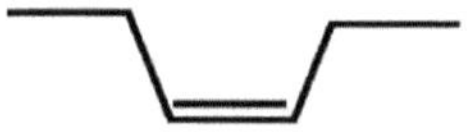 $KMnO_4$, KOH, H_2O, 15° →

(d) 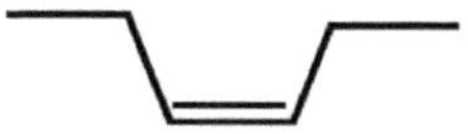 $KMnO_4$, KOH, H_2O, 85° →

(e) 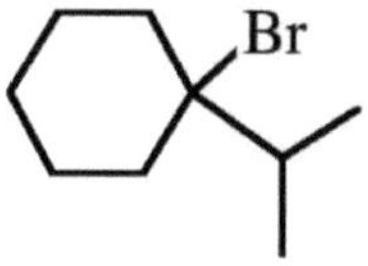

Br

$KOC(CH_3)_3$ →

3. (25 points) Give the correct reagents needed to accomplish each of the conversions indicated **E**-**J**.

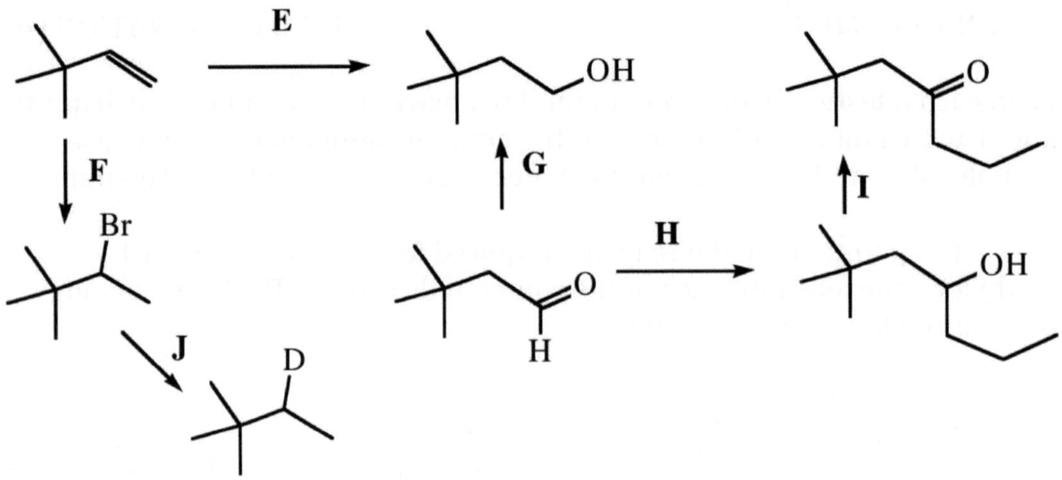

4. (25 points) Describe (tell exactly what you would *do* and what you would *see*) simple chemical tests that you would use to distinguish quickly between each of the following pairs of compounds:

(a) 3-ethyl-3-pentanol and 1-heptanol
(b) dibutyl ether and nonane
(c) 3-pentanone and pentanal

(d)

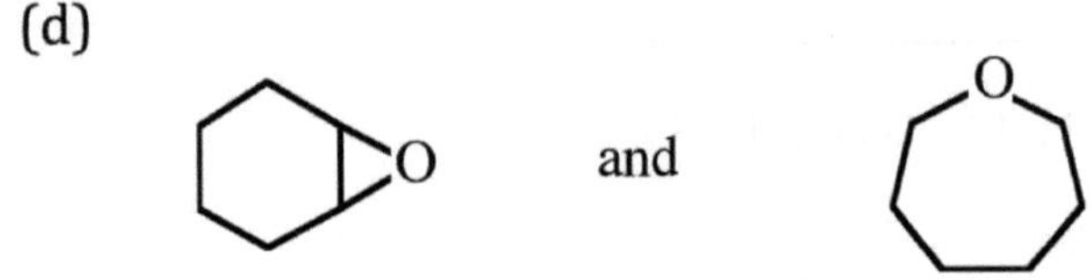

ORGANIC CHEMISTRY I FINAL EXAMINATION (I)

Read each question before you attempt to answer it. The value of each question is noted, with a total of 250 points for the entire examination. You may use a calculator or molecular models if you find them necessary. Time allowed = 150 minutes.

1. (20 points) Use the molecular orbital model to describe the C-I bond in H_3CI.

2. (26 points) Ethanol and hexane are miscible in all proportions with each other. Explain (give experimental details) how you could use water to separate the ethanol from the hexane.

3. (15 points) Consider the conformations of octane produced by rotation about the central carbon-carbon bond. Use Newman projections to depict each of the following:

 (a) the conformation of lowest energy
 (b) the conformation of highest energy
 (c) a staggered conformation intermediate in energy compared to the two previously drawn

4. (22 points) Use the curved arrow formalism to depict the mechanism of each of the following reactions:

 (a) the reaction of bromide ion with methyloxonium ion
 (b) the treatment of 2-methyl-2-propanol with sulfuric acid to give an alkene

5. (28 points) Give the structure of the major organic product formed from each of the following alcohols upon treatment with potassium permanganate in aqueous potassium hydroxide solution.

 (a) 2-methyl-1-butanol
 (b) 2-methyl-2-butanol
 (c) 2,4-dimethyl-3-hexanol
 (d) 2,2,6,6-tetramethyl-1-octanol

6. (28 points) For each of the following, complete the Fischer projection on the right so that it represents the same stereoisomer as the Newman projection on the left.

(a)

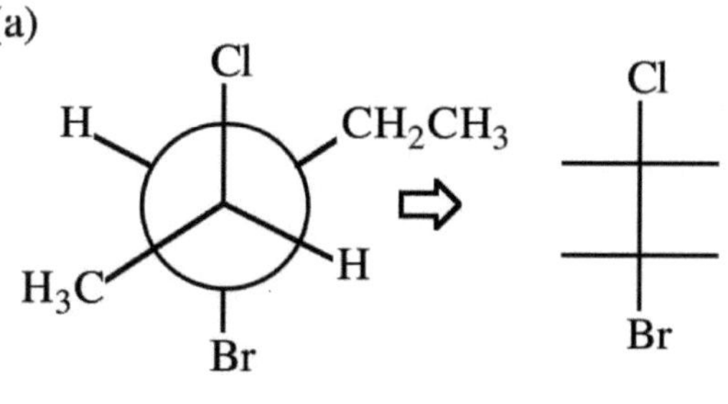

(b)

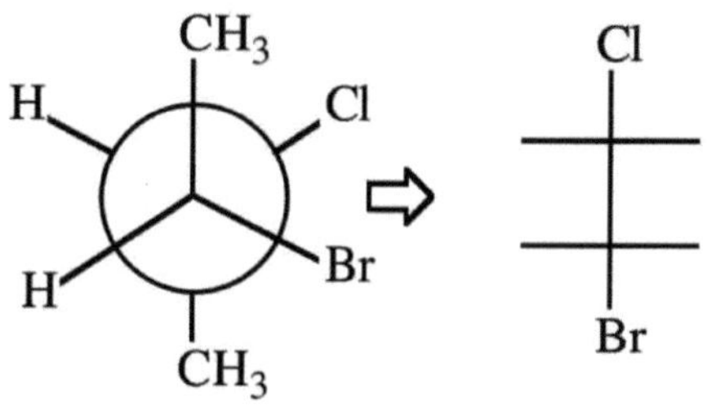

7. (25 points) A student plans to prepare 1-hexene by the acid-catalyzed dehydration of 1-hexanol. What problems do you see in accomplishing this synthesis? Devise a better synthetic route to 1-hexene from 1-hexanol.

8. (24 points) Give the best synthetic route for the preparation of each of the following target molecules from the indicated starting materials. You should give all reagents and reaction conditions. (More than one step may be required for the accomplishment of each synthesis.)

(a) $CH_3CH_2CO_2H$ from 1-butanol
(b) 1-pentanol from 3-bromopentane
(c) cyanocyclohexane from cyclohexanol
(d) (*S*)-2-cyanobutane from (*S*)-2-chlorobutane

9. (18 points) Give the major organic product showing it in its more stable chair conformation for each of the following reactions:

(a) cyclohexene treated with bromine in carbon tetrachloride solution
(b) 1-methylcyclohexene treated with BD_3 and worked up with D_2O/NaOD.

10. (24 points) Give the correct structure for each of the compounds indicated **A-D.**

A (C_3H_7Br) —Mg, ether→ [ ] —**B** (a ketone, $C_7H_{14}O$), aq. acid workup→ **C** ($C_{10}H_{22}O$) —H_2SO_4, heat→ **D** ($C_{10}H_{20}$)

D —1. O_3; 2. Zn, acetic acid→ H_3C–CO–CH_3 + $(H_3C)_2CH$–CO–$CH(CH_3)_2$

11. (20 points) Give the correct structure for each of the compounds indicated **E-G.** Show the correct stereochemistry where appropriate.

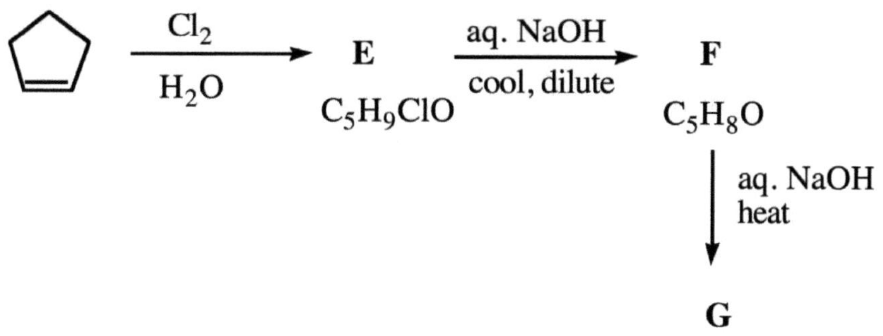

$\xrightarrow[H_2O]{Cl_2}$ **E** (C_5H_9ClO) $\xrightarrow[\text{cool, dilute}]{\text{aq. NaOH}}$ **F** (C_5H_8O)

aq. NaOH, heat ↓

G

Read each question before you attempt to answer it. Each question has a value of 25 points, with a total of 250 points for the entire examination. You may use a calculator or molecular models if you find them necessary. Time allowed = 150 minutes.

1. Using chemical methods, how would you distinguish between each of the following pairs of isomers? Tell what you would do and what you would see.

(a)  [cyclohexane] and [1-hexene]

(b) 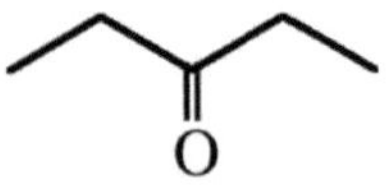 [3-pentanone] and [pentanal]

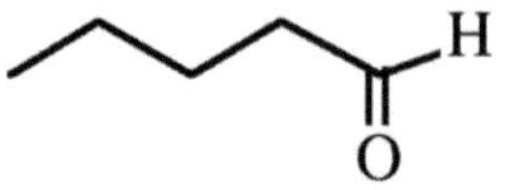

(c) 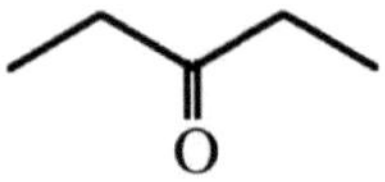 [3-pentanone] and [cyclopentanol]

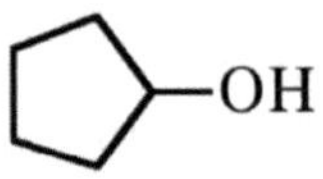

2. Give the structure of a compound that satisfies the criterion given in each case. There *may* be more than one correct answer for each.

(a) a six-carbon secondary alcohol that yields *only* a single alkene upon acid-catalyzed dehydration

(b) a six-carbon alkene that yields only a *meso*-dibromide upon reaction with Br_2/CCl_4

(c) a five-carbon chiral secondary alcohol that yields a five-carbon secondary alkyl chloride of (*S*) absolute configuration upon treatment with $SOCl_2$ in pyridine.

(d) a compound of formula $C_5H_{11}Br$ that can not yield an alkene upon treatment with potassium *tert*-butoxide.

3. Give the major organic product in each of the following reactions:

(a) cyclohexene treated with bromine in methanol

(b) 1-methylcyclohexene treated with "D_3B" followed by $H_2O_2/KOH/H_2O$

(c) cyclopentene heated with aqueous basic potassium permanganate

(d) (*E*)-3,4-dimethyl-3-hexene treated with ozone and worked up with Ni/CH_3CO_2H

(e) 1,2-dimethylcyclohexene treated with H_2/PtO_2.

4. Using the concept of molecular orbitals, explain why $(CH_3)_3C^+$ is more stable than is H_3C^+.

5. Consider the formation of $(CH_3)_3C\text{-}OH$ by the hydration of $(CH_3)_2C{=}CH_2$ in H_2O/ethanol with acid catalysis. Write a *mechanism* for the formation of the alcohol and draw a reaction progress diagram (energy *vs.* reaction coordinate) for the overall reaction.

6. Give the reagents required for each of the conversions **A-E**.

$(CH_3CH_2)_3C\text{-}H \xrightarrow{\textbf{A}} (CH_3CH_2)_3C\text{-}Br \xrightarrow{\textbf{B}} (CH_3CH_2)_2C{=}CH(CH_3)$

$(CH_3CH_2)_2C{=}CHCH_3 \xrightarrow{\textbf{C}} (CH_3CH_2)_2CH\text{-}CH(OH)CH_3$

$(CH_3CH_2)_2CH\text{-}CH(OH)CH_3 \xrightarrow{\textbf{D}} (CH_3CH_2)_2CH\text{-}C(=O)\text{-}CH_3$

$(CH_3CH_2)_2C{=}CHCH_3 \xrightarrow{\textbf{E}} (CH_3CH_2)_3C\text{-}OCH_3$

7. Give an explanation for the observation that *trans*-2-*tert*-butyl-1-chlorocyclohexane undergoes elimination reaction using potassium *tert*-butoxide much more slowly than does *cis*-2-*tert*-butyl-1-chlorocyclohexane.

8. A material of unknown structure that is found to contain only carbon, hydrogen, and oxygen, analyzes for 62.04% carbon and 10.41% hydrogen, is found to have a molecular weight of ~58, does not react with sodium metal, but does decolorize dilute aqueous basic potassium permanganate solution. Give a structure for this material based on the experimental data found.

9. The addition of HBr to an alkene in the presence of peroxides (R-O-O-R) is a free radical chain process. Illustrate the complete mechanism of the formation of an alkyl bromide by this route using $(CH_3)_2C{=}CH_2$ as the alkene.

10. Give reaction sequences which will accomplish each of the following syntheses using the indicated starting materials and any other reagents deemed necessary.

(a) synthesis of methyl 1-butyl ether from 2-butanol
(b) synthesis of 3-ethyl-3-pentanol from 3-pentanol
(c) synthesis of 2-deuteriopentane from 2-pentanol

Read each question before you attempt to answer it. Each question has a value of 25 points, with a total of 250 points for the entire examination. You may use a calculator or molecular models if you find them necessary. Time allowed = 150 minutes.

1. Give the structure of a compound that satisfies the criterion given in each instance. There *may* be more than one correct answer for each.
   (a) a ten-carbon alkene that yields *only* a single product upon ozonolysis with reductive work-up.
   (b) a seven-carbon alkene that yields only a racemic dibromide mixture upon reaction with Br_2/CCl_4.
   (c) a four-carbon alkene that yields a *meso*-diol upon treatment with formic acid/hydrogen peroxide/water.
   (d) a compound of formula $C_6H_{13}Br$ that yields only alkene upon treatment with sodium ethoxide.

2. Give the major organic product in each of the following reactions, being careful to illustrate the proper stereochemistry where appropriate:
   (a) cyclopentene treated with bromine in ethanol.
   (b) 1-ethylcyclopentene treated with $(CH_3)_2S{:}BD_3$ followed by CH_3COOH.
   (c) 3-methylpentane treated with Br_2 on irradiation with light to form a product of formula $C_6H_{13}Br$.
   (d) 1,2-dimethylcyclopentene treated with H_2 over PtO_2 catalyst.
   (e) (*R*)-2-iodoheptane allowed to stand for several days in a solution of NaI in acetone.

3. Using chemical methods, how would you distinguish between each of the following pairs of isomers? Tell what you would do and what you would see.

(a) 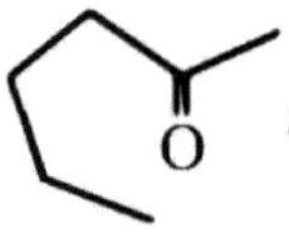 O and OH

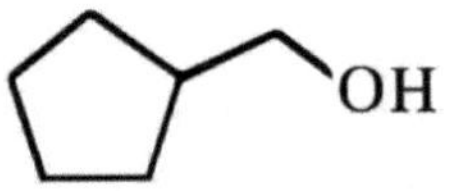

(b) 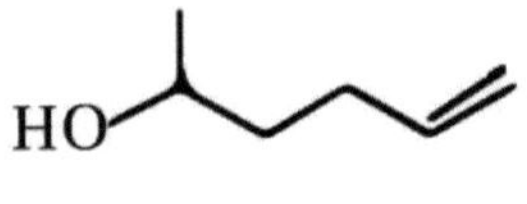 HO and OH

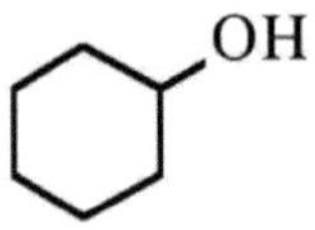

(c) 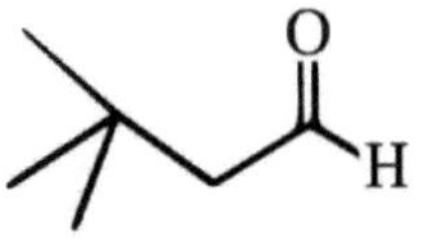 O H and OH

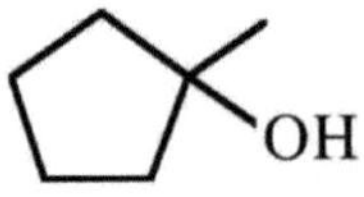

4. Using the concept of molecular orbitals, explain how CH_3O^- (methoxide ion) reacts with (*R*)-2-bromobutane to give (*S*)-2-butyl methyl ether.

5. Suppose that each of $(CH_3CH_2)_3CBr$ and $CH_3CH_2CH_2CH_2CH_2CH_2CH_2Br$ were to react with water, each following an S_N1 mechanism. Draw a reaction progress diagram (energy *vs.* reaction coordinate) for each reaction showing the relative ease or difficulty for each of the reactions to occur.

6. Give structures for each of the compounds indicated **A-E**.

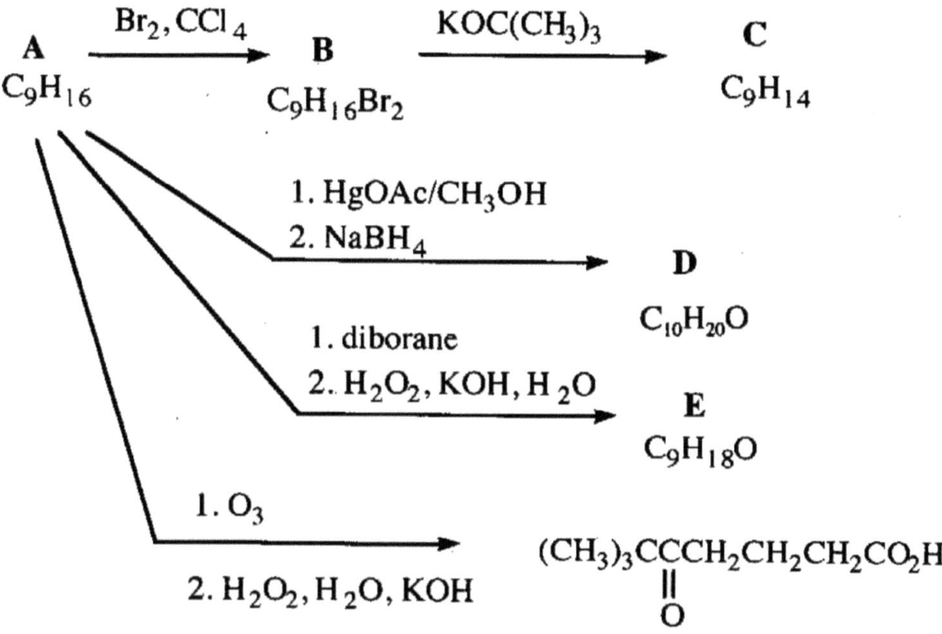

7. For each of the following, complete the stereochemical representation shown to the right such that it corresponds to the stereoisomer shown on the left.

(a)

(b)

8. Consider the reactions of *trans*-2-*tert*-butyl-1-bromocyclohexane and

cis-2-*tert*-butyl-1-bromocyclohexane with NaCN in dimethylsulfoxide solution (a hazardous reaction, but one that proceeds *via* the S_N2 mechanism). Which would react more quickly? Give an explanation of your choice.

9. Outline the most efficient synthesis of each of the following compounds from the indicated starting material using any other reagents deemed necessary:
   (a) 2-deuteriopentane from 2-pentanol
   (b) 3-ethyl-3-heptanol from 3-heptanol
   (c) $(CH_3)_3CCO_2H$ from $(CH_3)_3CCH_2CHBrC(CH_3)_3$

10. Give a concise definition for each of the following chemical terms:
   (a) resolution
   (b) nucleophile
   (c) LUMO
   (d) ß-elimination

Read each question before you attempt to answer it. The value of each question is noted, with a total of 100 points for the entire examination. You may use a calculator or molecular models if you find them necessary. Time allowed = 100 minutes.

1. (20 points) Give the correct structure for each of the compounds indicated **A**-**E**. Be careful to indicate the proper stereochemistry where appropriate.

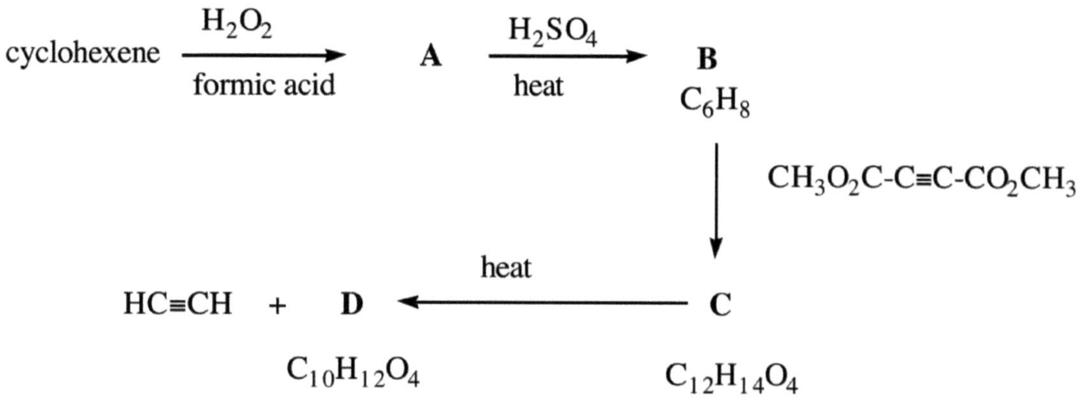

cyclohexene $\xrightarrow[\text{formic acid}]{H_2O_2}$ **A** $\xrightarrow[\text{heat}]{H_2SO_4}$ **B** (C_6H_8)

B $\xrightarrow{CH_3O_2C\text{-}C{\equiv}C\text{-}CO_2CH_3}$ **C** ($C_{12}H_{14}O_4$)

C $\xrightarrow{\text{heat}}$ HC≡CH + **D** ($C_{10}H_{12}O_4$)

2. (20 points) Give the correct reagent or set of reagents in order to accomplish each of the conversions shown below. More than one step may be required in each conversion.

(a) [alkene] ⟶ [ether with $OCH_2C_6H_5$]

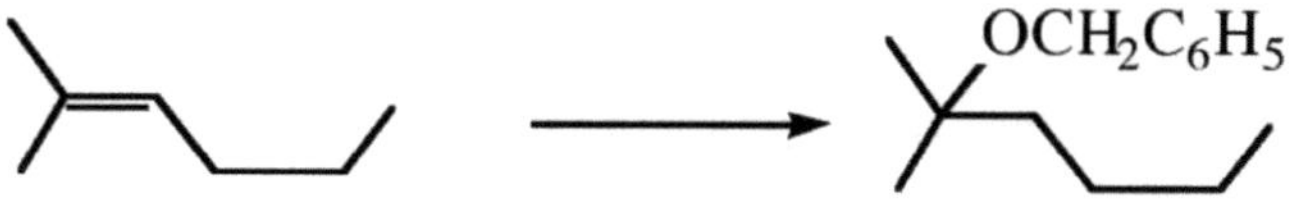

(b) [cyclohexyl alcohol, OH] ⟶ [$OCH_2C_6H_5$ ether]

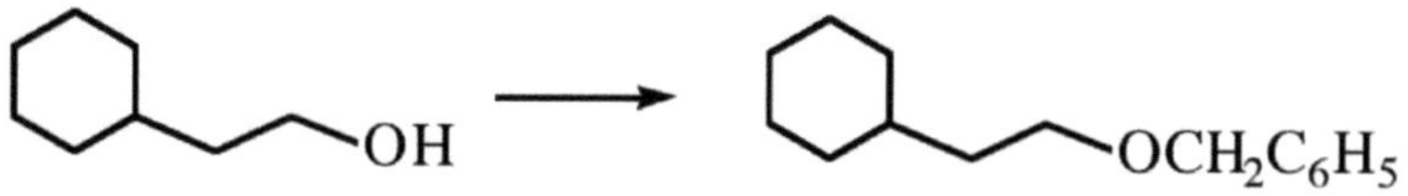

(c) [terminal alkyne] ⟶ [alkene]

(d) [$OCH_2C_6H_5$ ether] ⟶ [alcohol, OH]

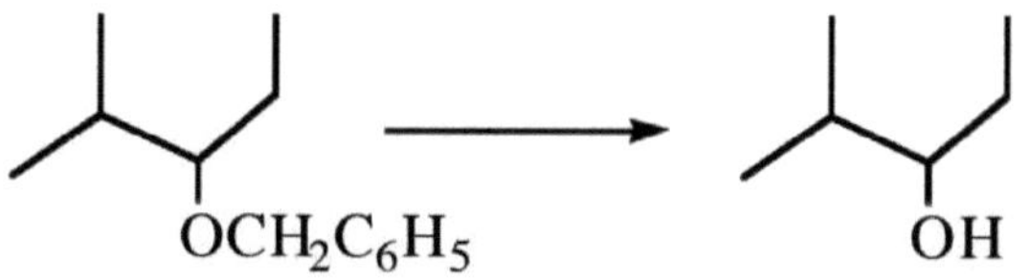

3. (20 points) Consider the pentadienyl anion. Sketch the orbital phases for the *filled* π molecular orbitals of this system. Does this anion have greater, lesser, or the same π stabilization compared to the cyclopentadienyl anion? Explain your answer.

4. (20 points) Predict the ^1H NMR spectrum (approximate chemical shift for the signal for each type of hydrogen, the relative integration of each signal, and the first-order splitting for each signal) for each of the following:

(b) $C_6H_5CH_2OCH_2$–CH=CH–$CH_2OCH_2C_6H_5$ in CCl_4 solution

(c) H, Br in D_2O solution

5. (20 points) Give the correct structure for each of the compounds indicated **F-J.**

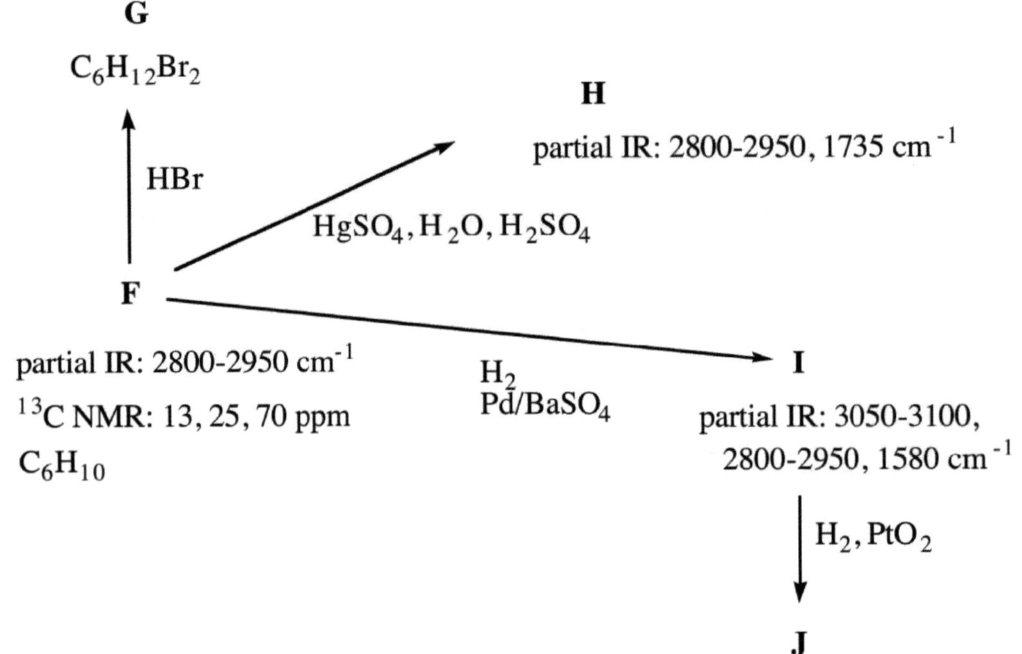

partial IR: 2800-2950 cm^{-1}

^{13}C NMR: 12, 17, 22 ppm

Read each question before you attempt to answer it. The value of each question is noted, with a total of 100 points for the entire examination. You may use a calculator or molecular models if you find them necessary. Time allowed = 100 minutes.

1. (20 points) Answer each of the following concerning molecular orbital systems:

(a) Sketch the π molecular orbitals for the filled molecular orbitals of the heptatrienyl cation [$(CH_2CHCHCHCHCHCH_2)^+$] indicating the phases of the individual lobes of each molecular orbital.

(b) Consider the transformation of these three orbitals for the construction of the corresponding cyclic π system (*i.e.* the filled molecular orbitals of the cycloheptatrienyl cation π system); show the relative stabilization or destabilization of each orbital upon this transformation.

(c) Consider the heptatrienyl anion [$(CH_2CHCHCHCHCHCH_2)^-$] and the heptatrienyl radical [$(CH_2CHCHCHCHCHCH_2)^{\cdot}$]; what is the *difference* in π bonding stabilization for these two species? Give a brief explanation.

2. (20 points) Give the major organic product in each of the following reactions. If more than one product is formed in significant amount, give all structures. Show the appropriate stereochemistry where appropriate. If no reaction occurs, so state.

(a)

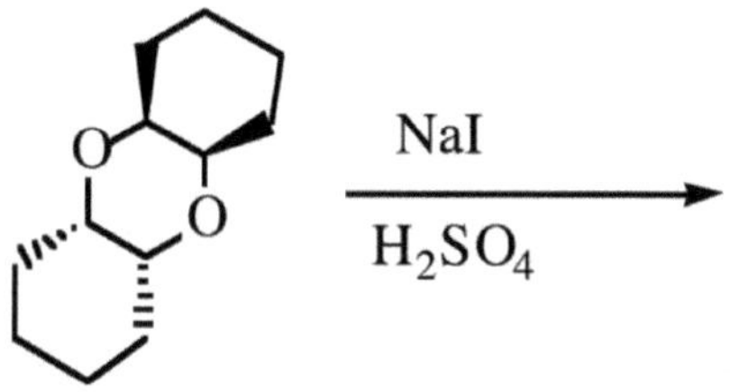

NaI, H_2SO_4 →

(b)

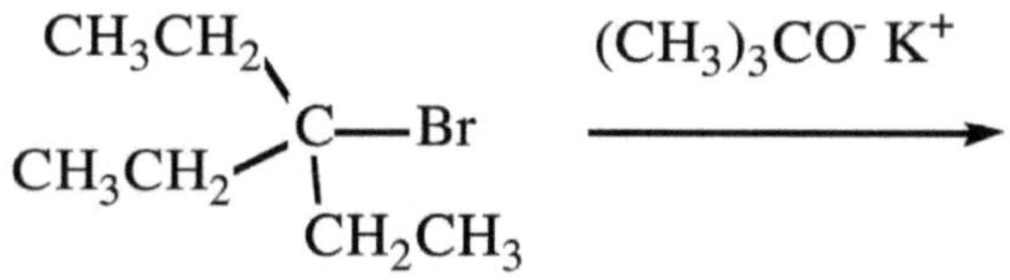

$(CH_3CH_2)_3C$—Br (drawn with three CH_3CH_2 groups and CH_2CH_3 on C) $\xrightarrow{(CH_3)_3CO^- K^+}$

(c)

HC≡C—cyclohexane $\xrightarrow[HgSO_4]{H_2O, H_2SO_4}$

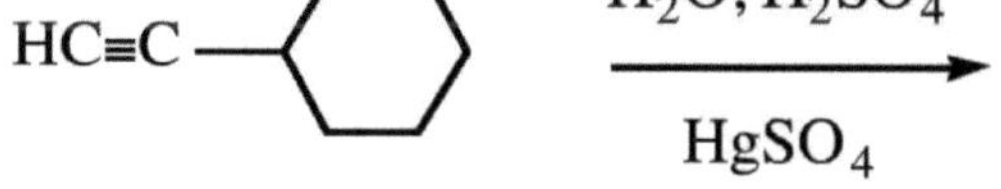

3. (18 points) Tell how you would distinguish between each of the following pairs of compounds using the indicated spectrometric technique.

(a)

$(CH_3)_2CH\text{-}O\text{-}CH(CH_3)_2$ and $(CH_3)_2CHCH(CH_3)_2$

(i) by ^1H NMR

(ii) by IR

(b)

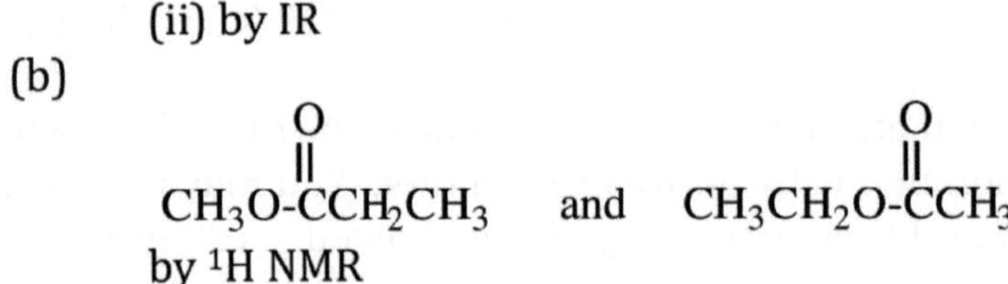

by ¹H NMR

4. (20 points) Give the correct structure for each of the compounds indicated **A-E.**

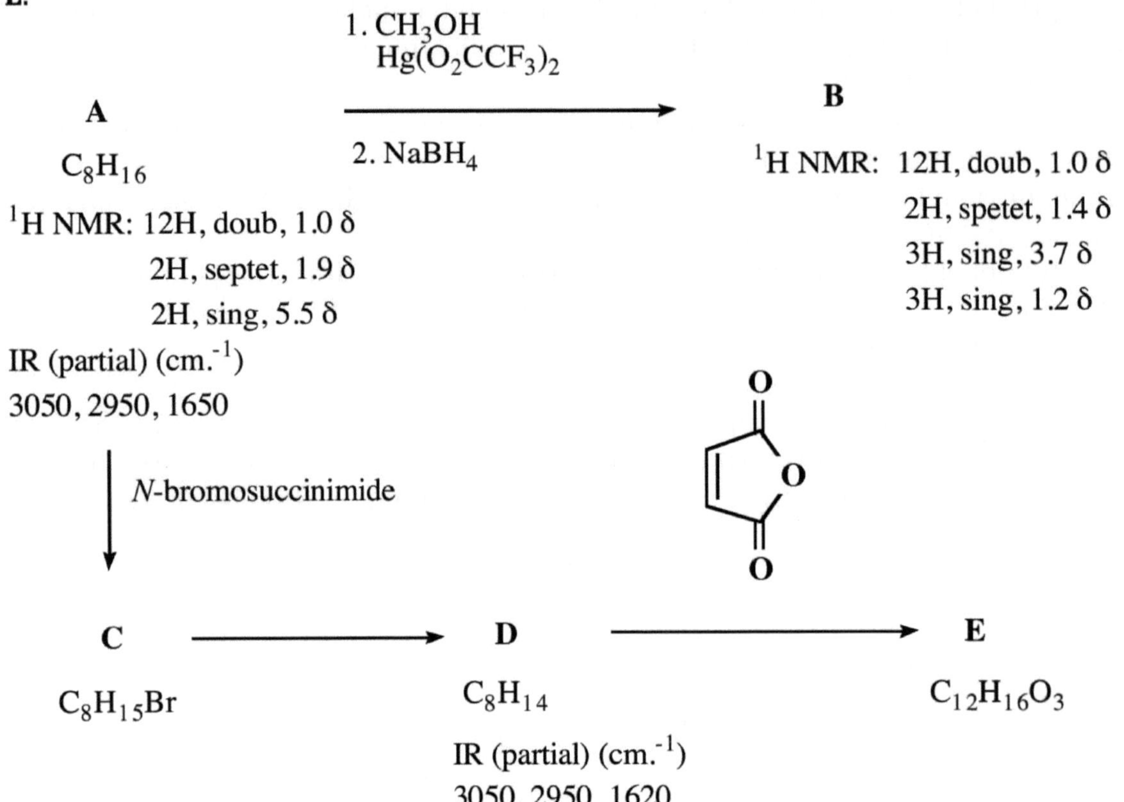

5. (22 points) Give the best synthesis of each of the following compounds from the indicated starting materials and any other reagents deemed necessary.

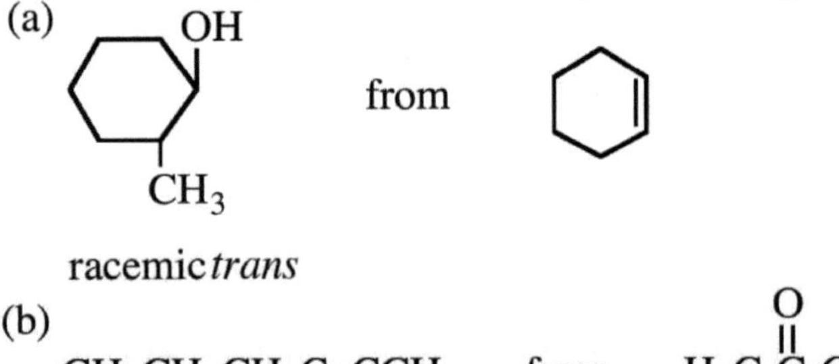

(b)

$CH_3CH_2CH_2C{\equiv}CCH_3$ from $H_3C\text{-}\overset{O}{\overset{\|}{C}}\text{-}CH_3$

(c)

(4*R*,5*S*)-4,5-dibromoöctane from $CH_3CH_2CH_2C{\equiv}CCH_2CH_2CH_3$

Read each question before you attempt to answer it. The value of each question is noted, with a total of 100 points for the entire examination. You may use a calculator or molecular models if you find them necessary. Time allowed = 100 minutes.

1. (15 points) Give the correct structure for each of the compounds indicated **A-C**.

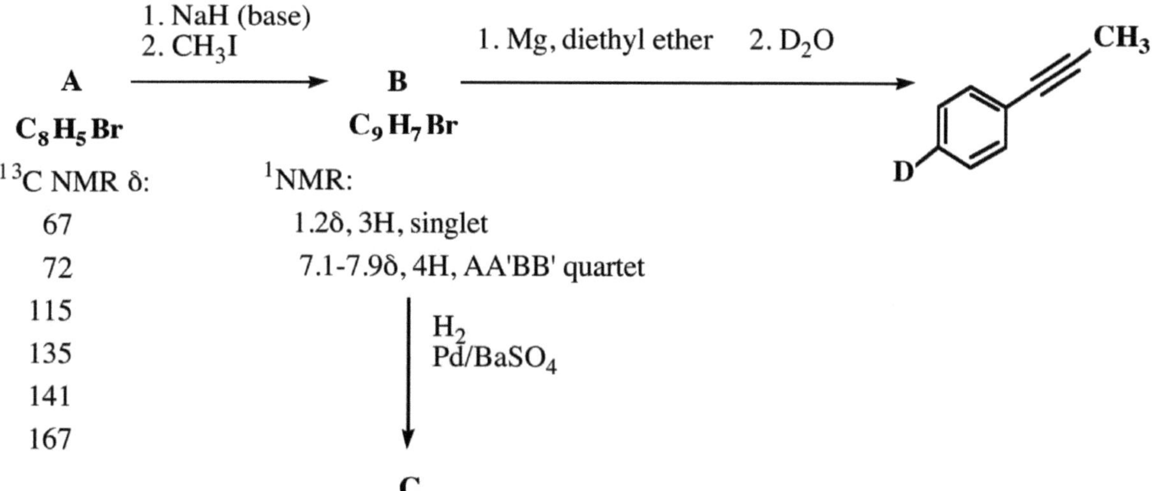

2. (10 points) Give the major organic product in each of the following reactions, indicating correct stereochemistry where appropriate. If more than one product is formed in significant amount, give all structures. If no reaction occurs, so state.

(a)

CH_3 $\xrightarrow{H_2,\ Pd/BaSO_4}$

(b)

\+ O O O $\xrightarrow{\text{heat}}$

3. (14 points) Give the reagents required for each of the following conversions. More than one step may be required in each instance. You may use any other organic or inorganic reagents deemed necessary.

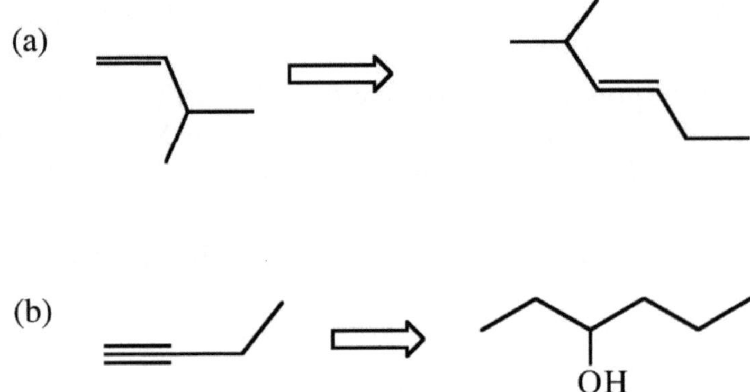

4. (20 points) Consider the pentadienyl π-electronic system. Draw the energy diagram for the *pentadienyl cation*, indicating the proper number of electrons in the proper orbitals for the ground state of the cation. Also, draw the orbital pictures for the HOMO and the LUMO of this pentadienyl cation. For the addition of two more electrons to this π-electronic system, what would be the resultant change in π-bonding stabilization from the cationic system you have already considered? Explain your answer.

5. (20 points) For each of the following species, concisely explain *why* it is not aromatic.

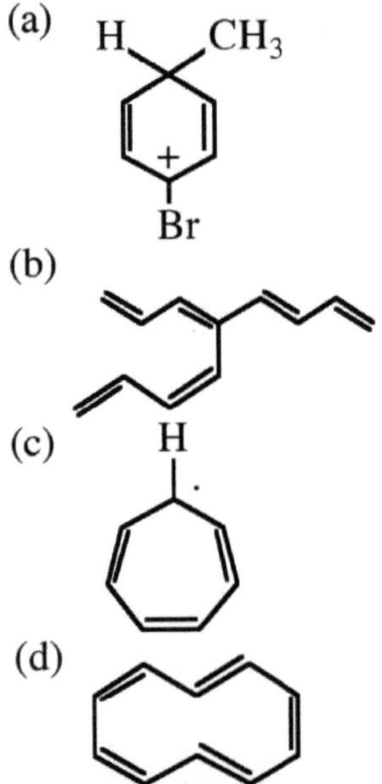

6. (21 points) Predict the ^{1}H NMR spectrum (approximate chemical shift for the signal for each type of hydrogen, the relative integration of each signal, and the first-order splitting for each signal) for each of the following:

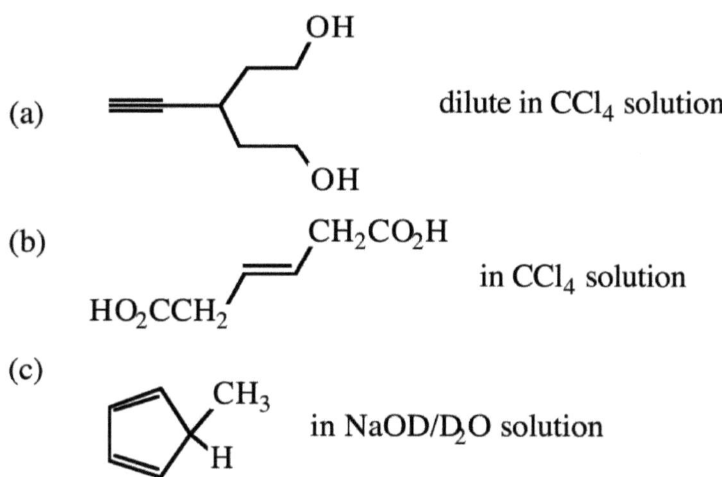
(a)
OH
OH
dilute in CCl_4 solution
(b)
CH_2CO_2H
HO_2CCH_2
in CCl_4 solution
(c)
CH_3
H
in NaOD/D_2O solution

Read each question before you attempt to answer it. The value of each question is noted, with a total of 100 points for the entire examination. You may use a calculator or molecular models if you find them necessary. Time allowed = 100 minutes.

1. (25 points) Give the *complete* mechanism for the hydrolysis, under aqueous acidic conditions, of the iminium salt as shown in the overall reaction below. You should show all steps in the overall reaction including the structures of all of the intermediate species.

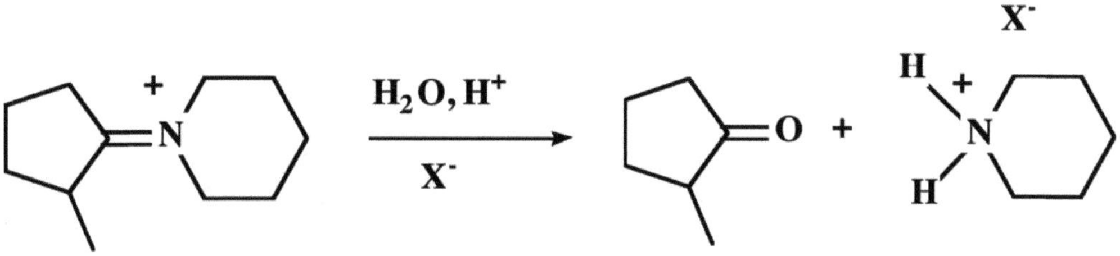

H_2O, H^+

X^-

X^-

H

N

H

2. (27 points) Give complete synthetic procedures for the preparation of each of the following compounds starting with either diethyl malonate or ethyl acetoacetate as the source of all carbon atoms of the products, and using any other organic and inorganic reagents deemed necessary.

(a)

CO_2H

(b)

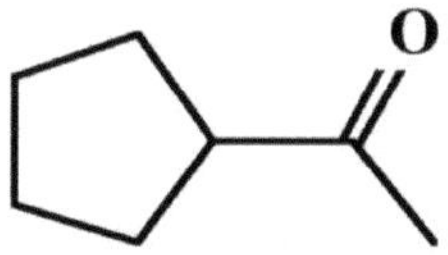

O

(c)

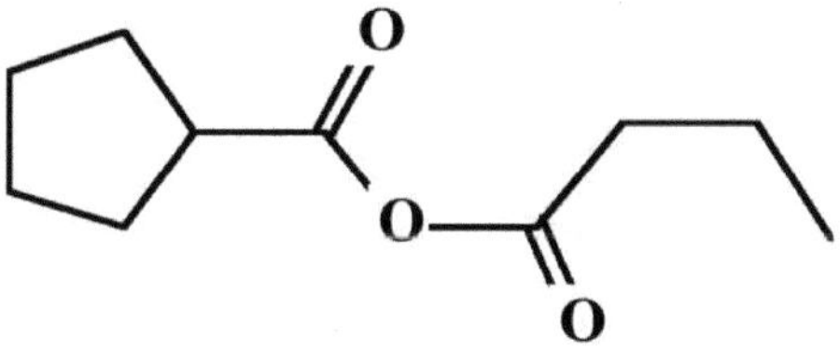

O

O

O

3. (20 points) Consider the overall conversion shown immediately below.

HO–(CH$_2$)$_4$–C(=O)OCH$_3$ ⟶ (δ-valerolactone) + CH_3OH

Describe in detail the reaction conditions that would be required for the formation of the product lactone in good yield as well as the difficulties that would be encountered in the performance of the reaction.

4. (28 points) Give the correct structures for each of the compounds indicated **A-G.**

A $C_6H_{11}O_2Ag$ —(Br_2, CCl_4)→ **B** $C_5H_{11}Br$

A: ^1H NMR: 0.9 δ, 9H, singlet; 2.1 δ, 2H, singlet

B: ^1H NMR: 0.9 δ, 9H, singlet; 3.5 δ, 2H, singlet

B —(1. potassium phthalimide; 2. aq. acid)→ **C** $C_5H_{13}N$

C + $CH_3CH_2CH_2CH_2CHO$ —(acid catalyst)→ $CH_3CH_2CH_2CH_2CH{=}N{-}CH_2C(CH_3)_3$

C —(1. $H_2C{=}O$; 2. HCO_2H)→ **D** $C_6H_{15}N$

D —(excess CH_3I)→ **F** $C_8H_{20}NI$

F —(H_2O, KOH, heat)→ **No Reaction**

D —($CH_3C(=O)OC(=O)CH_3$)→ **E** $C_8H_{17}NO$

E —(LiAlH$_4$, ether; aq. acid workup)→ **G** $C_8H_{19}N$

Read each question before you attempt to answer it. The value of each question is noted, with a total of 100 points for the entire examination. You may use a calculator or molecular models if you find them necessary. Time allowed = 100 minutes.

1. (20 points) Give the reagents that would be required for each of the following conversions. More than one step may be needed in each overall conversion. You may use any other organic or inorganic reagents deemed necessary.

(a) Br ⇨ CO_2H

(b) CO_2H ⇨ Br

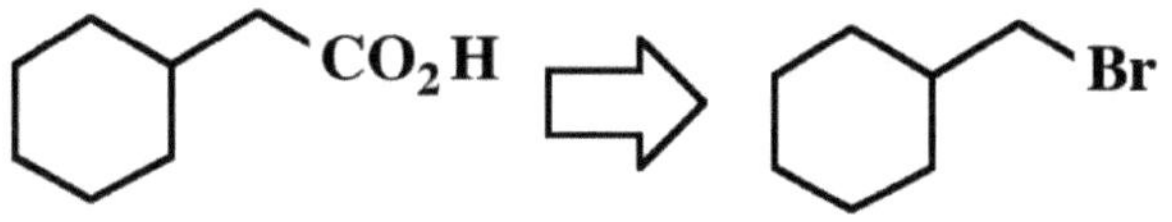

(c) O ⇨ Br CO_2H

(d) ⇨ Br Cl

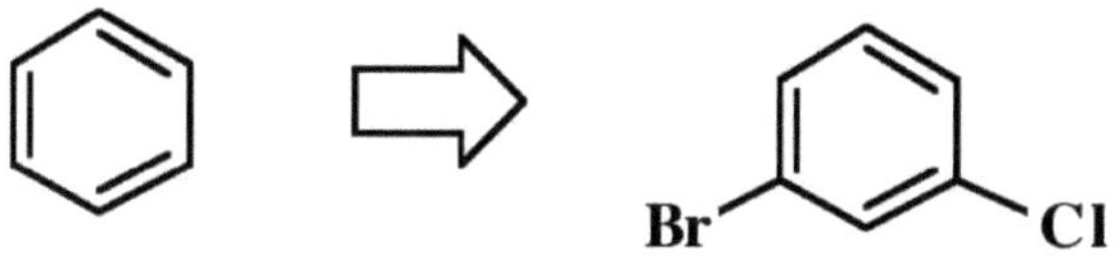

2. (20 points) Give the *complete* mechanism for the Fischer esterification as shown below.

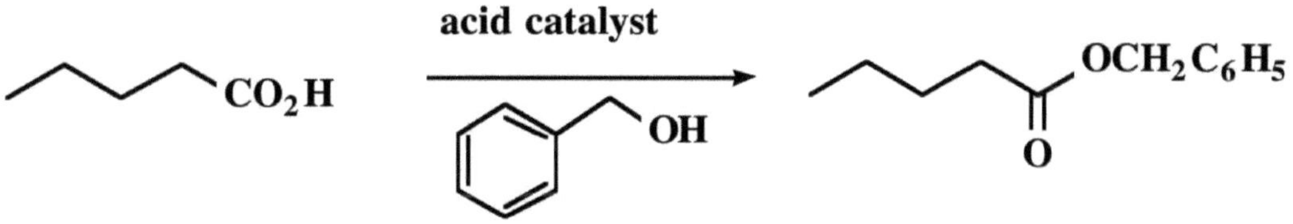

CO_2H —acid catalyst, OH→ $OCH_2C_6H_5$ O

3. (15 points) Predict the relative order of acidities for the series of compounds shown below. Indicate their relative acidities with "1" being the least acidic to "5" being the most acidic.

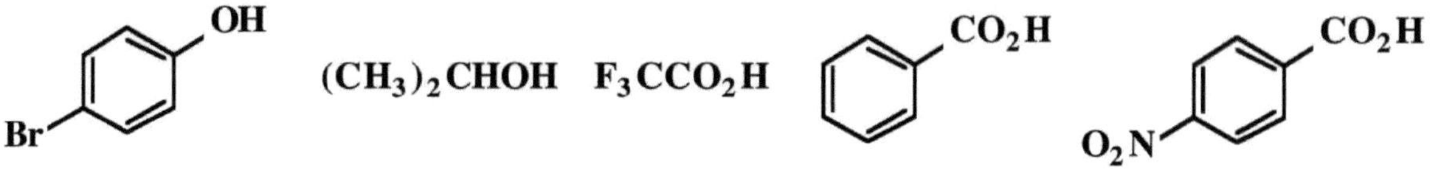

OH Br $(CH_3)_2CHOH$ F_3CCO_2H CO_2H CO_2H O_2N

4. (15 points) In each instance, starting with heptanoic acid, give the reagents and reaction condition required for its conversion into each of the compounds listed below. You may use any other inorganic reagents and organic compounds deemed necessary, but heptanoic acid must be the starting point of each synthesis.

(a) 2-bromoheptanoic acid
(b) 1-bromohexane
(c) decane

5. (30 points) Give the correct structures for each of the compounds indicated **A-H** in the following reaction scheme.

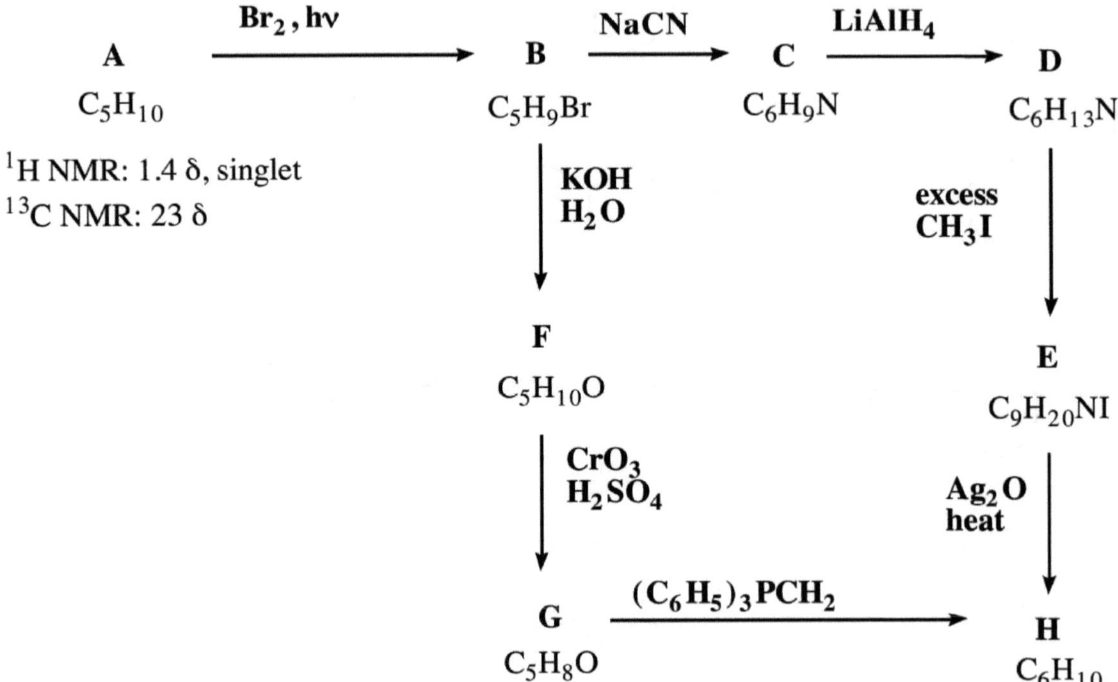

Read each question before you attempt to answer it. The value of each question is noted, with a total of 100 points for the entire examination. You may use a calculator or molecular models if you find them necessary. Time allowed = 100 minutes.

1. (20 points) Give the reagents that would be required for each of the following conversions. More than one step may be needed in each overall conversion. You may use any other organic or inorganic reagents deemed necessary.

(a) Br ⇨ CO_2H

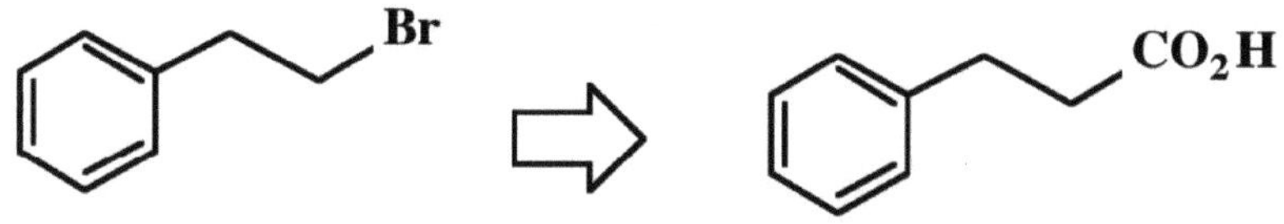

(b) ⇨ Cl Cl

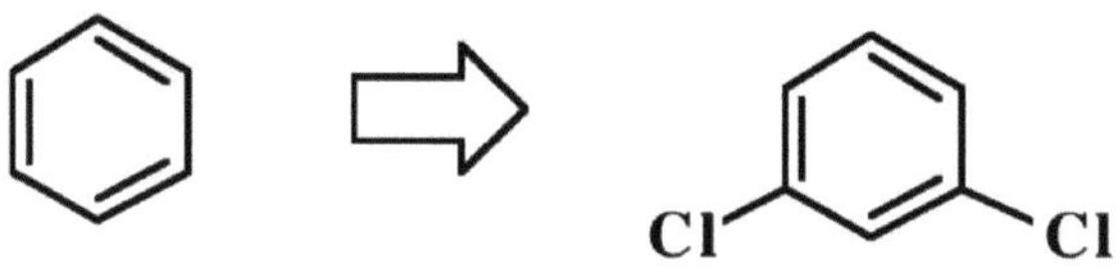

(c) NH ⇨ N-$CH_2C_6H_5$

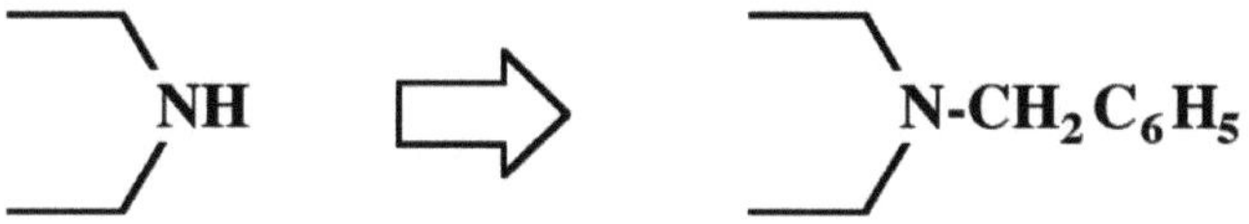

(d) O ⇨ Br

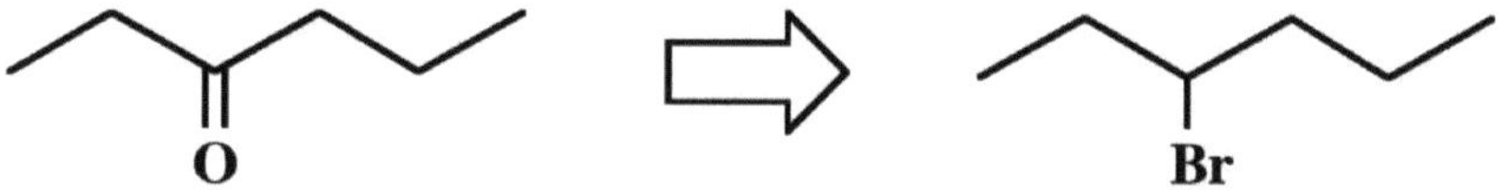

2. (20 points) Give the *complete* mechanism for the hydrolysis reaction shown below. You should show all steps in the overall conversion, including the structures for all intermediate species.

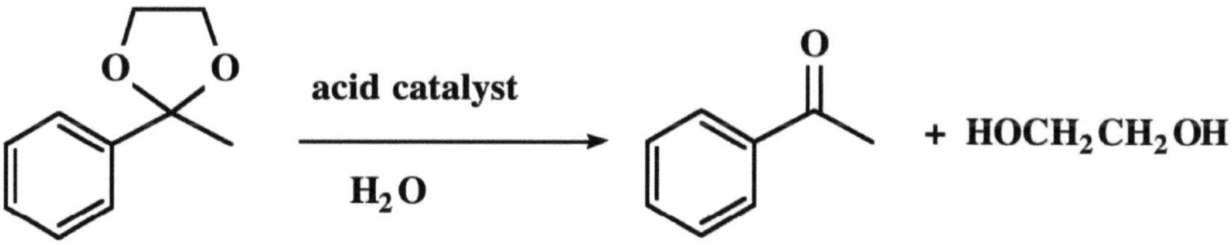

O O —acid catalyst / H_2O→ O + $HOCH_2CH_2OH$

3. (20 points) give the correct structure for each of the compounds indicated **A-E.**

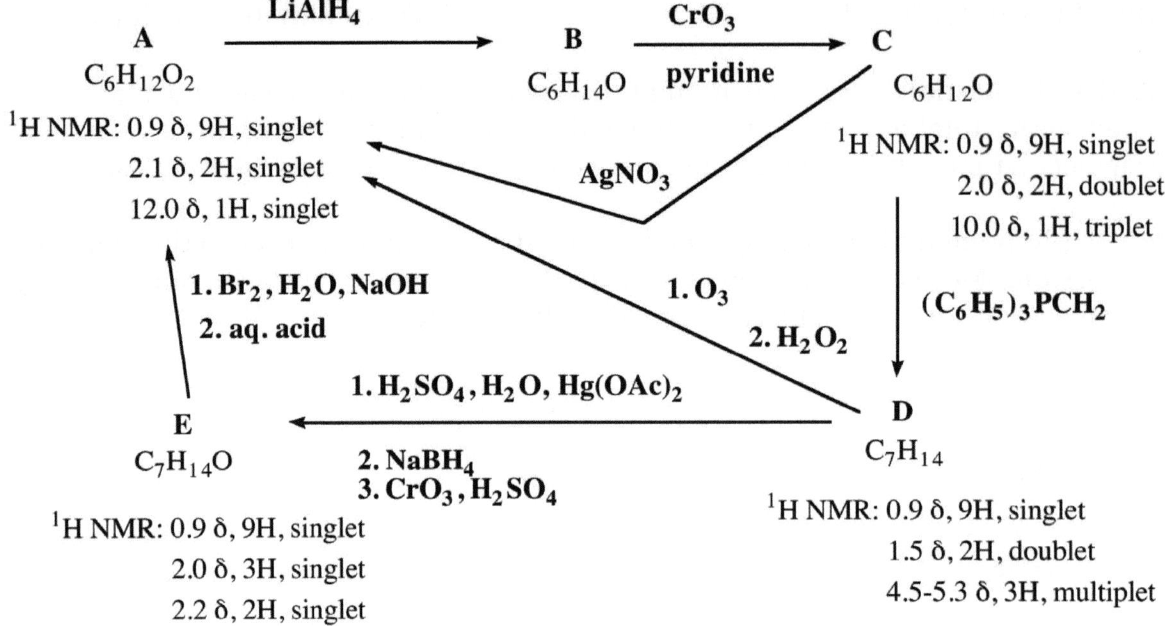

4. (20 points) Predict the relative order of basicities for the series of compounds shown below. List them from most basic to least basic, and give a brief explanation for the position of each compound in the series.

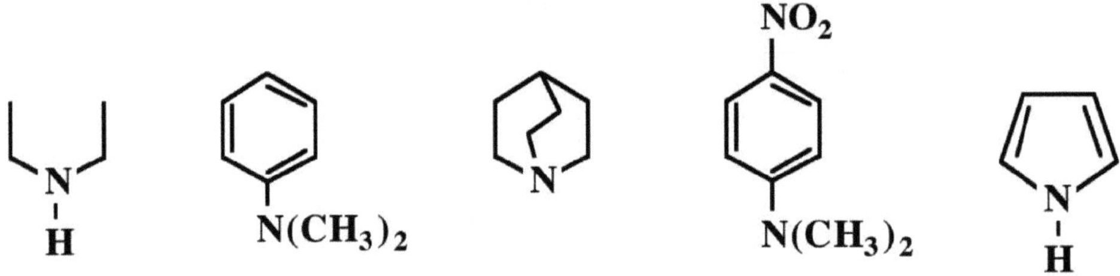

5. (20 points) Predict the major organic product in each of the following reactions:

(a)

1. $KMnO_4, KOH, H_2O$

2. aq. acid

(b)

O

O_2N

$Sn(Hg), HCl, H_2O$

(c)

NH_2

+

O

acid

(d)

$N(CH_3)_2$

$NaNO_2, H_2O$

H_2SO_4

ORGANIC CHEMISTRY II THIRD EXAMINATION (I)

Read each question before you attempt to answer it. The value of each question is noted, with a total of 100 points for the entire examination. You may use a calculator or molecular models if you find them necessary. Time allowed = 100 minutes.

1. (20 points) Provide the correct reagents and structures **I-VI** for the reaction scheme shown below. Be sure to show the proper stereochemistry where required.

HO
OH
D-mannitol diacetonide
I
II
H O
H CH_2
OsO_4
III + IV
$(C_7H_{14}O_4)$ $(C_7H_{14}O_4)$
aq. acid
aq. acid
V
$(C_4H_{10}O_4)$
optically active
VI
$(C_4H_{10}O_4)$
optically inactive

2. (20 points) You are synthesizing alanylglycine starting with alanine and glycine. Explain why:

(a) you need to protect the amino nitrogen of the alanine;
(b) you need to protect the carboxyl group of glycine;
(c) isobutyl chloroformate can serve to activate the coupling process; and
(d) protecting the amino nitrogen using acetyl chloride is *not* a good idea.

3. (20 points) You are assigned by your supervisor the task of synthesizing, in the most economical possible manner, 2-ethylcyclohexanone. Justify to your supervisor using a longer route, proceeding through the enamine shown below,

which requires the purchase of piperidine, an additional reagent, rather than simply treating cyclohexanone with sodium hydroxide and ethyl iodide.

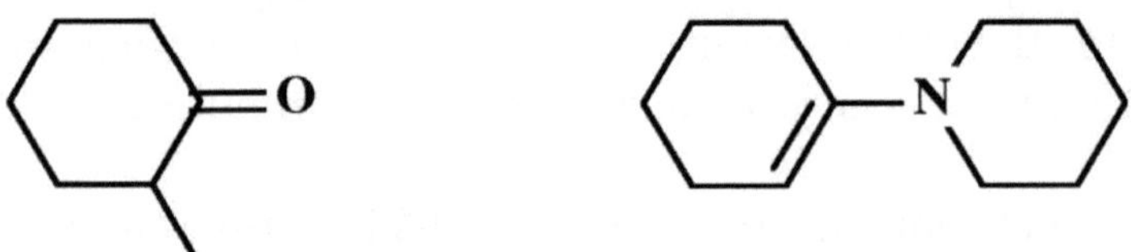

2-ethylcyclohexanone

N--1-cyclohexenylpiperidine (an enamine)

4. (20 points) Show a complete synthesis of racemic phenylalanine starting with ethyl acetoacetate and using any other organic and inorganic reagents deemed necessary. At least two carbon atoms of the original ethyl acetoacetate must be retained in the final phenylalanine product.

5. (20 points) Give the correct structure of the major organic product in each of the following reactions:

(a) $H_2C{=}CH\text{-}CHO + NaCH(CO_2CH_3)_2 \xrightarrow{\text{water work-up}}$

(b) $(CH_3CO)_2O + NaO_2CCH_3 + C_6H_5CHO \xrightarrow{\text{water work-up}}$

(c) $CH_3O_2C(CH_2)_6CO_2CH_3 + NaOCH_3 \longrightarrow$

(d) $H_2C{=}CH\text{-}CHO + AgNO_3 \xrightarrow{\text{aqueous ammonia}}$

Read each question before you attempt to answer it. The value of each question is noted, with a total of 100 points for the entire examination. You may use a calculator or molecular models if you find them necessary. Time allowed = 100 minutes.

1. (20 points) Write out in detail the complete mechanism for the laboratory preparation of the enamine from cyclohexanone and pyrrolidine in the presence of an acid catalyst. The overall reaction is indicated below.

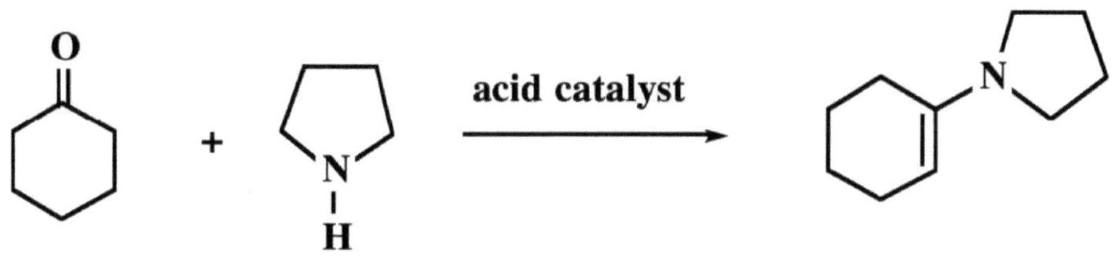

O + N H acid catalyst N

2. (15 points) An organic chemistry student is assigned to prepare 5 g (expecting 80% isolated yield) of the compound shown below as his laboratory final examination project. He mixes all of then reagents shown in the scheme simultaneously, and ultimately isolates only 0.4 g of the desired product. Explain what went wrong. Describe how the preparation should have been performed.

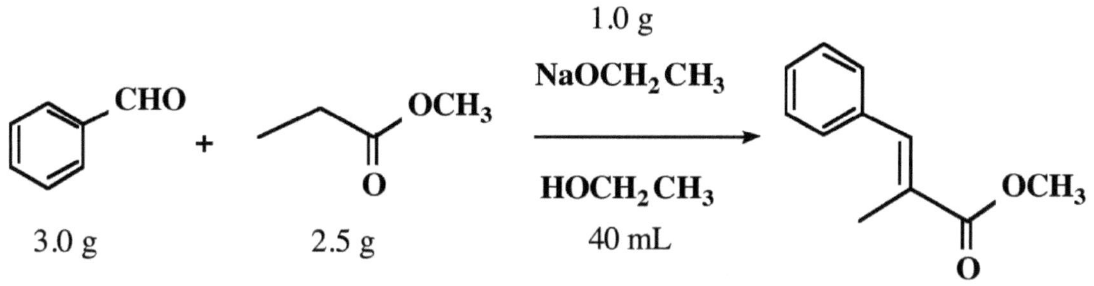

CHO (3.0 g) + OCH$_3$ O (2.5 g) — 1.0 g NaOCH$_2$CH$_3$ / HOCH$_2$CH$_3$ 40 mL → OCH$_3$ O

3. (24 points) You have isolated from a plant a carbohydrate material (elementary analysis fits the empirical formula $C_nH_{2n}O_n$) for which you have determined a molecular weight of 120. The ^{13}C NMR exhibits four signals (40, 55, 62 and 204 δ). Describe the experiments you would need to perform to establish the complete structure of this material, including stereochemistry, assuming that the only carbohydrate with which you could make any comparisons is D-glyceraldehyde.

4. (21 points) Give the complete synthesis for each of the following compounds beginning with the indicated organic starting material and using any other reagents (organic or inorganic) deemed necessary.

(a)

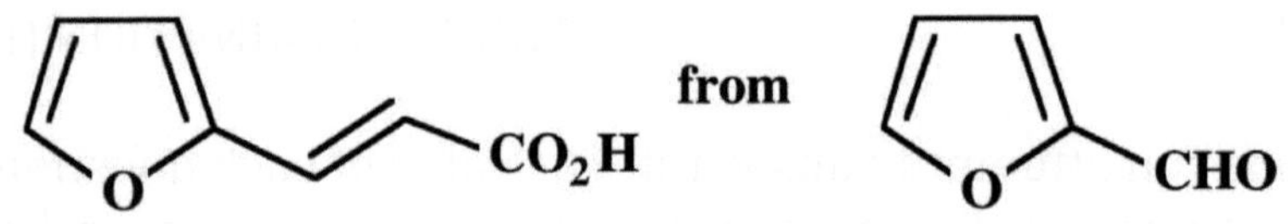

from CO_2H ... CHO

(b)

NH_2 from $CONH_2$

(c)

$CH_2CH_2NH_2$ from $CONH_2$

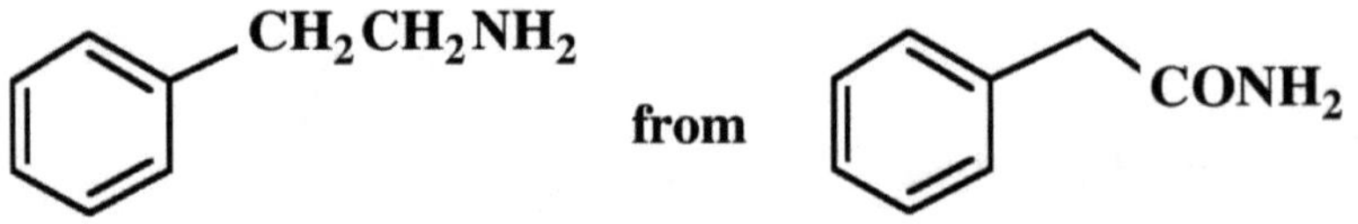

5. (20 points) Give a molecular orbital rationalization for the addition of enolate anion ($H_2C{=}CHO^-$) at the β-carbon of acrolein to form $HOCH_2CH_2C(O)H$ rather than at the carbonyl carbon.

Read each question before you attempt to answer it. The value of each question is noted, with a total of 100 points for the entire examination. You may use a calculator or molecular models if you find them necessary. Time allowed = 100 minutes.

1. (24 points) Provide structures for each of the compounds indicated **A-F**.

A ($C_6H_{12}O_6$) $\xrightarrow[(CH_3)_2CHOH]{NaBH_4}$ **B** ($C_6H_{14}O_6$) $\xrightarrow[acid]{(CH_3)_2C{=}O}$ **C** ($C_{12}H_{22}O_6$)

C $\xrightarrow{NaIO_4}$ 2 **D** (2 equivalents of **D** from 1 equivalent of **C**) ($C_6H_{10}O_3$)

2 **D** $\xrightarrow[(CH_3)_2CHOH]{NaBH_4}$ **E** ($C_6H_{12}O_3$) $\xrightarrow{CH_3C(O)Cl}$ **F** ($C_8H_{14}O_4$)

F $\xrightarrow{H_3O^+}$

CH_2OH
H—C—OH
$CH_2OC(O)CH_3$

2. (18 points) Suppose we prepare alanine by the series of reactions illustrated below.

$CH_3CH_2CO_2CH_3$ $\xrightarrow[acetic\ acid]{Br_2}$ $CH_3CHBrCO_2CH_3$

$CH_3CHBrCO_2CH_3$ $\xrightarrow{\text{potassium phthalimide (NK)}}$ N-phthalimido-$CH(CH_3)CO_2CH_3$ (phthalimide $NCH(CH_3)CO_2CH_3$)

$NCH(CH_3)CO_2CH_3$ (phthalimide) $\xrightarrow[aq.\ acid]{heat}$ $CH_3CH(NH_2)CO_2H$ (alanine)

If one mole of the alanine thus produced is treated with one mole of D-tartaric acid (structure shown below) in methanol solution at room temperature, two products are formed which have significantly different solubilities in the methanol. Give structures for the two products and explain why they would exhibit different solubilities in methanol.

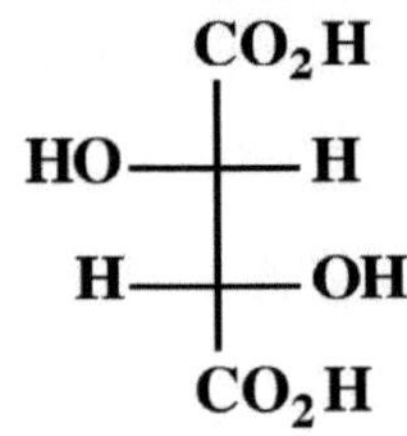

3. (27 points) Give synthetic procedures (indicate all reagents and reaction conditions) for the preparation of phenylalanine from each of the following starting compounds:

(a) glycine
(b) 3-bromopropanoic acid
(c) phenylacetaldehyde

4. (19 points) Concisely outline the procedure that would be used to synthesize L-alanyl-L-phenylalanine starting with the parent amino acids.

5. (12 points) Give concise definitions for each of the following terms:

(a) molecular ion
(b) molar absorptivity
(c) vacuum ultraviolet

ORGANIC CHEMISTRY II FINAL EXAMINATION (I)

Read each question before you attempt to answer it. The value of each question is noted, with a total of 250 points for the entire examination. You may use a calculator or molecular models if you find them necessary. Time allowed = 150 minutes.

1. (25 points) Give the best synthesis for each of the following compounds starting with alcohols containing five (5) or fewer carbon atoms as the only source of carbon in the products, and using any other inorganic or organic reagents deemed necessary.

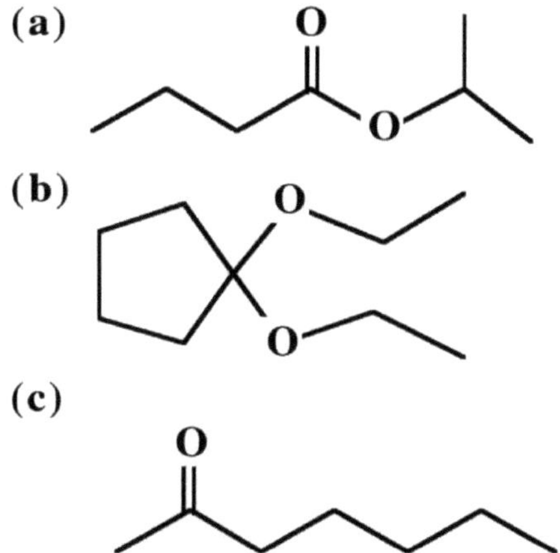

2. (30 points) Give the correct structure for each of the compounds indicated **A-F** and the reagent **G** required for the direct conversion of **B** to **F**.

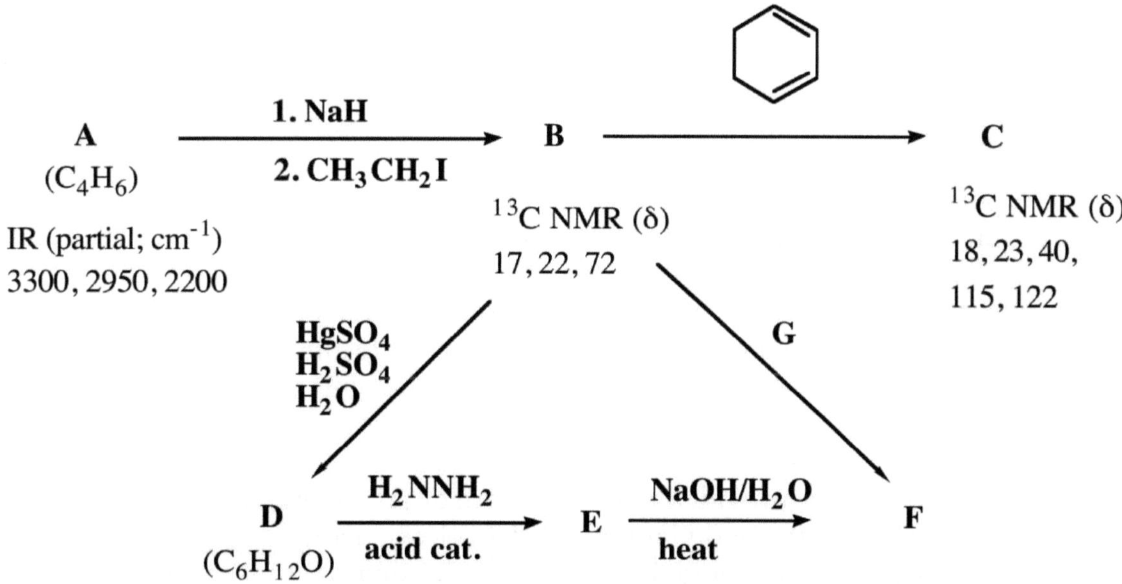

3. (30 points) Give the best set of reagents that would be used to accomplish each of the following conversions. More than one step may be used in each conversion.

(a) [benzene ring]-NO_2 to Br-[benzene ring]-CO_2H

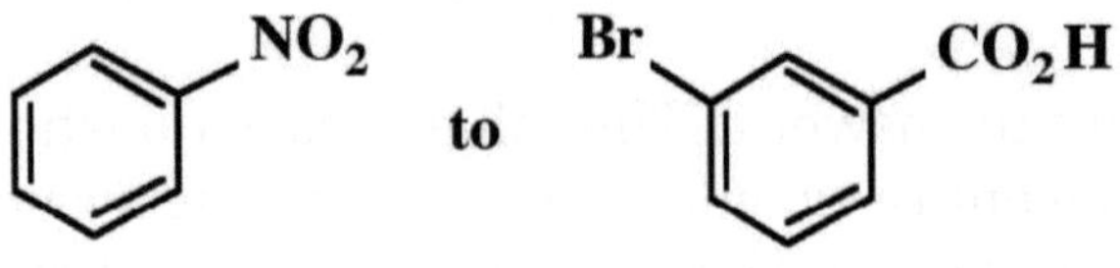

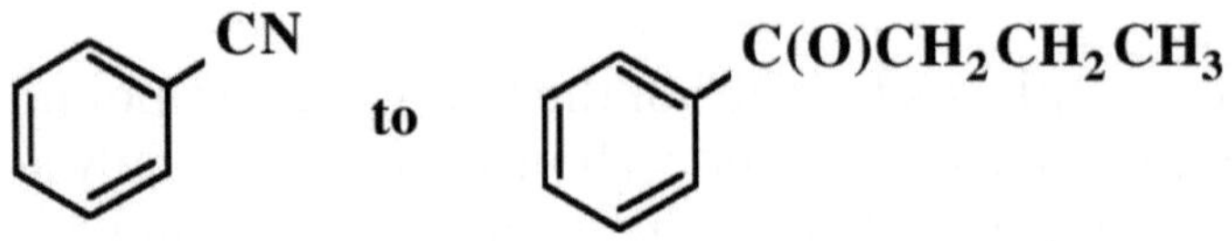

(b) [benzene ring]-CN to [benzene ring]-$C(O)CH_2CH_2CH_3$

(c) [benzene ring] to H_3C-[benzene ring]-$C(O)CH_2CH_3$

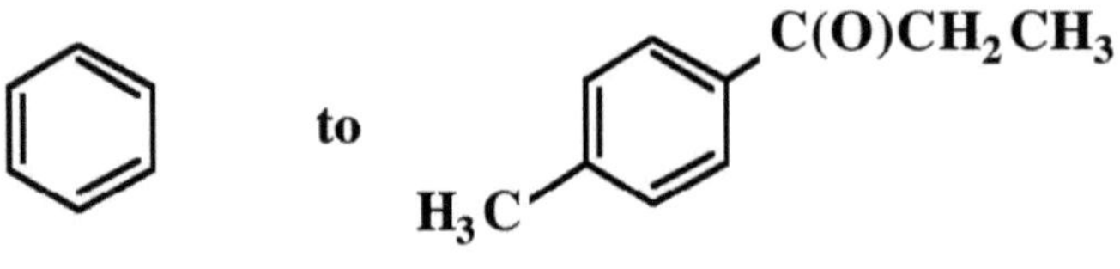

4. (25 points) Outline in detail the reactions and procedures you would use in determining the sequence of amino acids in a tripeptide that has been analyzed and found to contain one unit each of phenylalanine, alanine, and glycine. You should show structures of the reaction products for which you would be analyzing.

5. (30 points) Give a complete mechanism for the overall process of formation of a phenylhydrazone, as shown below.

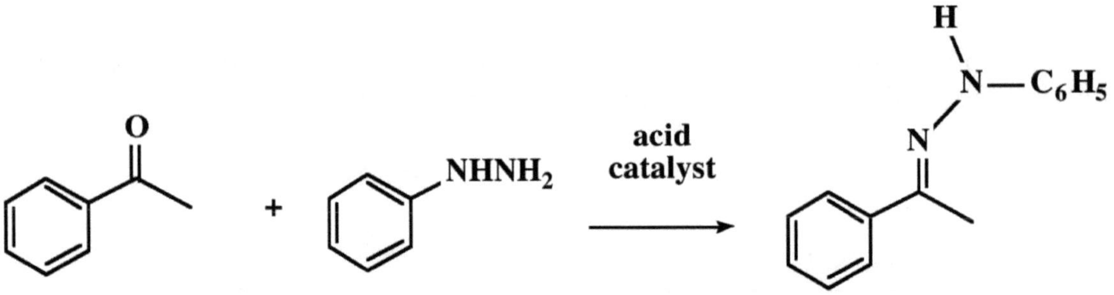

[acetophenone (O)] + [benzene ring]-$NHNH_2$ $\xrightarrow[\text{catalyst}]{\text{acid}}$ [benzene ring]-C(=N-N(H)-C_6H_5)-CH_3

6. (25 points) Consider the concerted cycloaddition reaction involving 1,3-butadiene and 1,3,5,7-octatetraene to form 1,3,5,9-cyclododecatetraene. Draw out the π molecular orbitals that you would need to consider when looking at the feasibility of this cycloaddition, and, using these molecular orbitals, explain: (1) if the reaction would proceed most feasibly under thermal or photochemical stimulation, and (2) if any other cycloaddition reaction might reasonably compete with it.

7. (30 points) You are provided with a mixture of the following three compounds: ethyl benzoate, 4-ethylaniline, and 4-ethylbenzoic acid. Give the experimental procedure you would follow in order to separate these three compounds from each other into pure samples of each.

8. (25 points) You are working in an analytical laboratory and the only tool you have available is an infra-red spectrometer. You are given three samples in bottles

from which the labels have fallen off, and it's your job to decide which label goes with which bottle. You know that the three samples are: (1) 4-ethoxybenzoic acid; (2) ethyl benzoate; (3) 4-ethylbenzoic acid. Tell how you would be able to decide which label went with which bottle simply on the basis of infra-red spectra that you measure.

9. (30 points) Given the ^{1}H and ^{13}C NMR spectral data along with other information provided, deduce the structure for each of the following compounds:

(a) molecular weight 116; ^{1}H NMR: 1.9 δ, 3H, singlet; 8.3 δ, 5H, broad singlet; ^{13}C NMR (δ) 25, 67, 72, 111, 120, 131, 140.

(b) formula $C_8H_8Br_2$; ^{1}H NMR: 3.7 δ, 4H, singlet; 8.1 δ, 4H, singlet; ^{13}C NMR (δ) 64, 121, 130.

Read each question before you attempt to answer it. The value of each question is noted, with a total of 250 points for the entire examination. You may use a calculator or molecular models if you find them necessary. Time allowed = 150 minutes.

1. (20 points) Explain how you would distinguish between each of the following pairs of isomers using UV spectrometry.

(a)

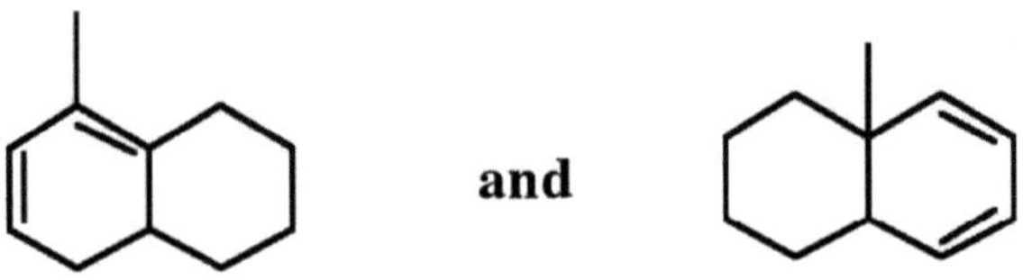

and

(b)

O

and

O

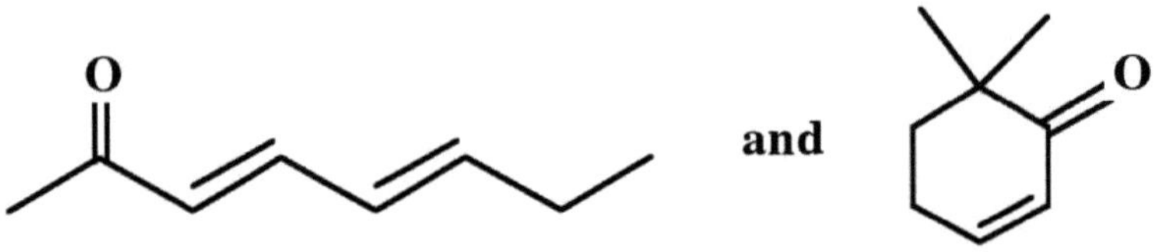

2. (30 points) Compound **A** is a D-aldopentose that can be oxidized to an optically inactive aldaric acid, **B.** On Kiliani-Fischer chain extension **A** is converted into **C** and **D**. The compound **C** can be oxidized to an optically inactive aldaric acid, **E**, but **D** is oxidized to an optically active aldaric acid, **F**. Give the structures for compounds **A-F**.

3. (32 points) Give the best set of reagents and reaction conitions for the preparation of each of the following from the indicated starting materials. You may use any other organic or inorganic reagents deemed necessary. More than one step may be required in each synthesis.

(a) $CH_3C(O)CH(CH_3)CO_2CH_2CH_3$ from $CH_3CO_2CH_2CH_3$

(b) [m-nitrobenzoic acid: benzene ring with CO_2H and NO_2] from [benzene]

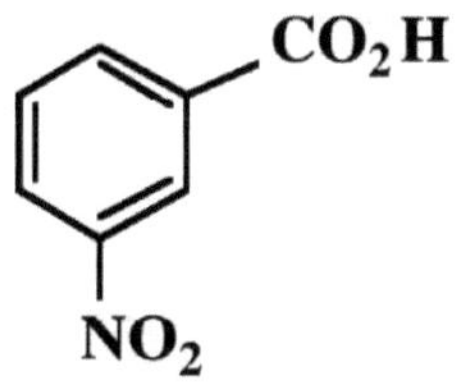

(c) [benzene ring with $C(O)CH_2CH_3$] from $CH_3CH_2CO_2H$

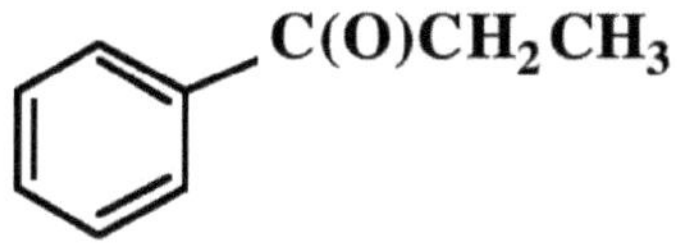

(d) [cyclohexane ring with exocyclic double bond to C(CH₃)CH₂CH₃] from $CH_3CH_2CH(Br)CH_3$

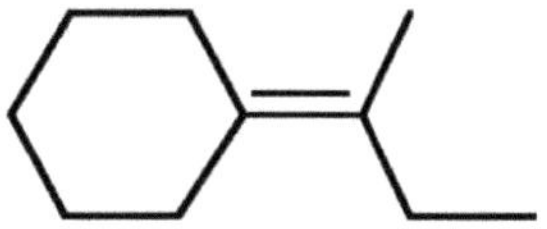

4. (20 points) Which would you anticipate to be the stronger acid, cyclopropene or cyclopentadiene? Explain your answer.

5. (28 points) Give structures for each of the compounds **G-I**.

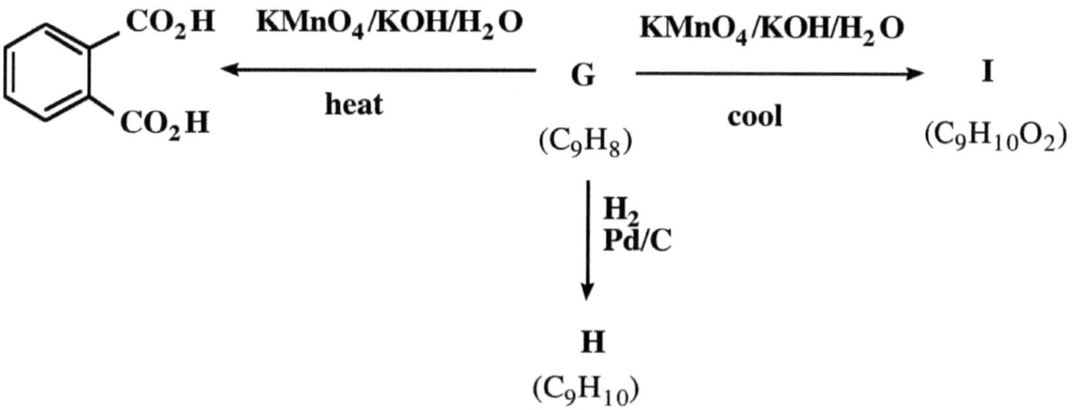

[phthalic acid: benzene ring with two CO_2H] $\xleftarrow[\text{heat}]{KMnO_4/KOH/H_2O}$ G (C_9H_8) $\xrightarrow[\text{cool}]{KMnO_4/KOH/H_2O}$ I ($C_9H_{10}O_2$)

G $\xrightarrow{H_2,\ Pd/C}$ H

H

(C_9H_{10})

^{1}H NMR:

quintet, 2.04 δ, relative area 1

triplet, 2.91 δ, relative area 2

singlet, 7.17 δ, relative area 2

6. (25 points) Give a complete mechanism for the hydrolysis of the ketal as shown below.

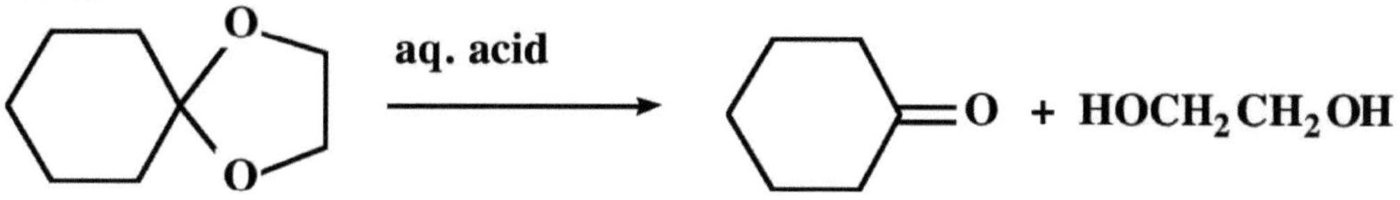

[cyclohexanone ethylene ketal] $\xrightarrow{\text{aq. acid}}$ [cyclohexanone]=O + $HOCH_2CH_2OH$

7. (25 points) Compound **J** exhibits a molecular ion in the mass spectrum of m/e = 122 with an (M+1) peak that is 8.9% as intense as the molecular ion peak. The IR spectrum (CCl_4) exhibits the following absorptions (among others - all in cm^{-1}): 3300 (broad), 3040, 2980, 1210. The ^{1}H NMR spectrum (D_2O) exhibits the following signals (δ): 2.4, 2H, triplet; 3.6, 2H, triplet; 8.1, 5H, broad singlet.

8. (20 points) Perform a retrosynthetic analysis to discover two possible synthetic routes for preparing 2-methyl-2-hexanol from a carbonyl compound and a Grignard reagent.

9. (30 points) Describe how you would synthesize each of the following compounds starting with either benzene or toluene.

   (a) 4-bromoaniline
   (b) benzylamine
   (c) 3-bromoaniline

10. (20 points) Give the major organic product that is formed in each of the following reactions:

   (a) the heating of benzonitrile with hydrochloric acid
   (b) the treatment of benzonitrile with the Grignard reagent derived from bromobenzene
   (c) the treatment of ethanal with ammonia and hydrogen cyanide
   (d) the treatment of ethanal with diethylamine and formaldehyde in the presence of an acid catalyst

Answer all questions in the answer book provided to you. Each of the ten regular questions is worth 25 points for a total of 250 points.

1. Contemplate the molecules 1,3-butadiene and 1,3,5-hexatriene. For the ground state molecules, tell which would exhibit the smaller HOMO-LUMO gap and explain how the HOMO-LUMO gap would change as conjugation is extended to larger fully-conjugated polyenes.

2. Give clear, concise definitions for each of the following terms:
   a) glycolysis
   b) Hofmann orientation
   c) aldohexose
   d) deactivating group for electrophilic aromatic substitution

3. Give the correct structure for each of the compounds indicated **A-E** in the following reaction scheme, being sure to show the proper stereochemistry, where relevant.

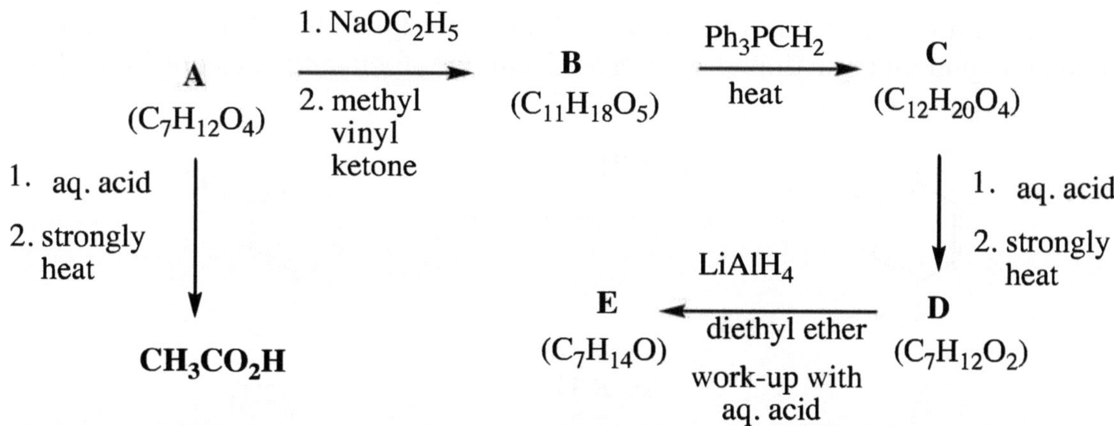

4. Provide an example of each of the following types of reactions using real reactants and products:
   a) Wittig olefination
   b) Fischer esterification
   c) Enamine monomethylation of a ketone
   d) Ketal formation

5. Provide the reagents and reaction conditions for each of the following conversions. More than one step may be required in each conversion.

a)

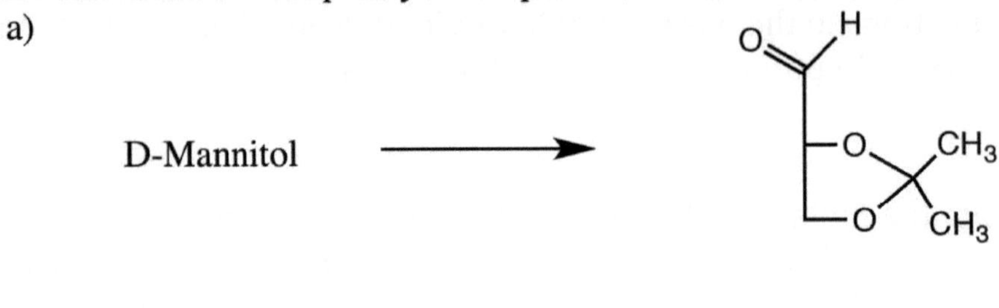

b) Glycine → racemic Phenylalanine

c) 1-Bromopropane → Ethyl butanoate

d) Benzene → *m*-Dichlorobenzene

6. The reagent dicyclohexylcarbodiimide in the presence of a trace of strong acid (*e.g.* H_3PO_4) can be used to form *N*-butyl benzamide from 1-butylamine and benzoic acid. Give a complete mechanism for this reaction.

7. The material 4-chloro-1,3-dinitrobenzene undergoes an unusual reaction for an aromatic compound, as illustrated below. Concisely explain how this reaction occurs but a corresponding reaction of 4-chloro-1,3-dimethylbenzene does not (also shown below as a non-reaction).

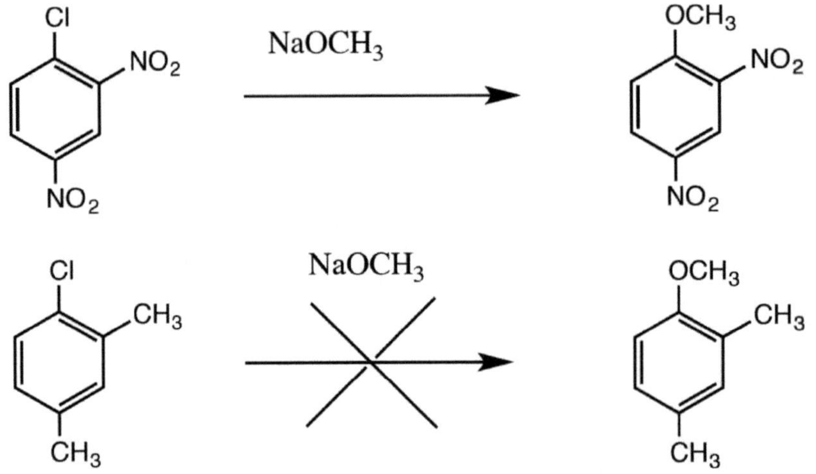

8. Predict the major organic product in each of the following reactions, showing the proper stereochemistry where appropriate:

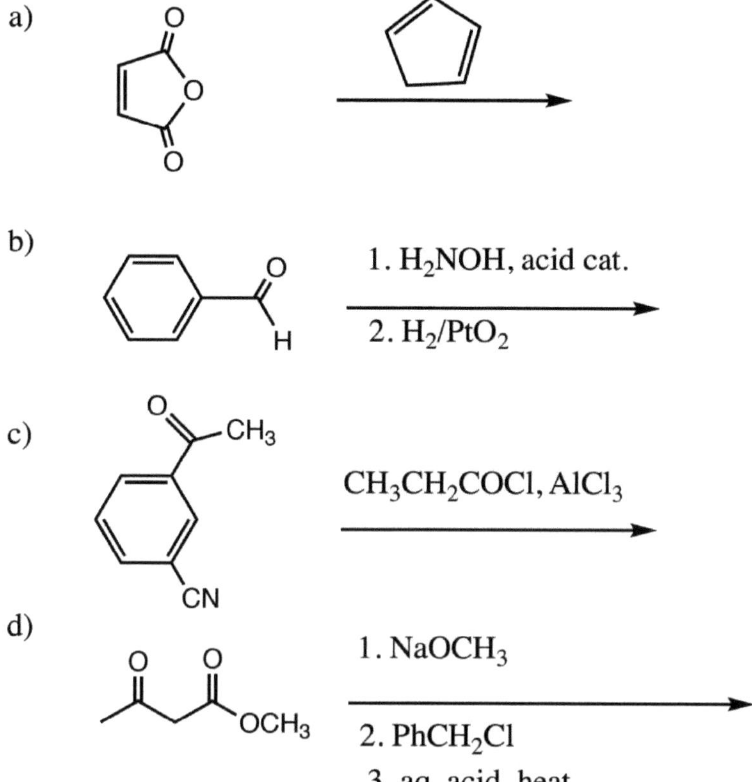

9. Trichloroacetonitrile ($Cl_3C\text{-}CN$) can serve as a coupling agent for ester formation, as in the conversion shown below. Give a mechanism for the way trichloroacetonitrile can so serve.

$$CH_3CH_2CO_2H \xrightarrow[Cl_3C\text{-}CN]{HOCH_2CH_2CH_2CH_3} CH_3CH_2CO_2CH_2CH_2CH_2CH_3 \quad [+ Cl_3C(O)NH_2]$$

10. Show the reagents involved in each of the steps of the conversion of D-ribose to D-erythrose.

ANSWERS

ORGANIC CHEMISTRY I — FIRST EXAMINATION (I)

ANSWERS

1. (10 points)

(a)

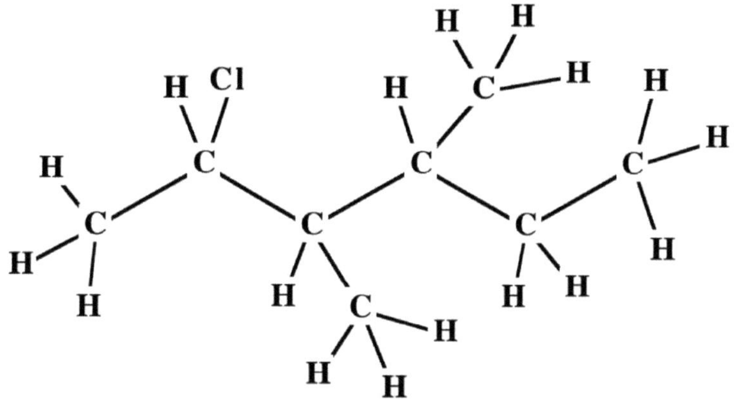

(b)

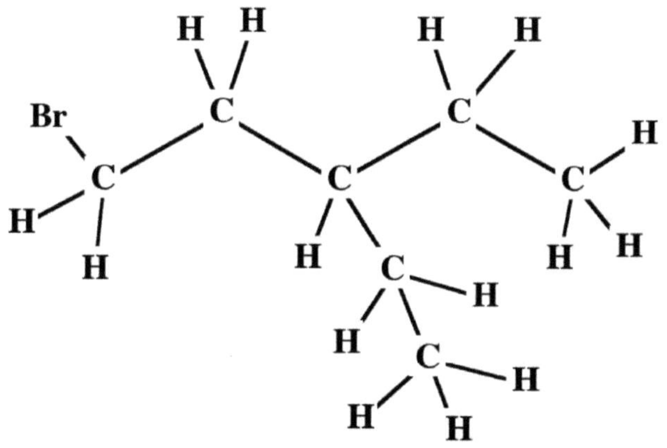

2. (20 points)

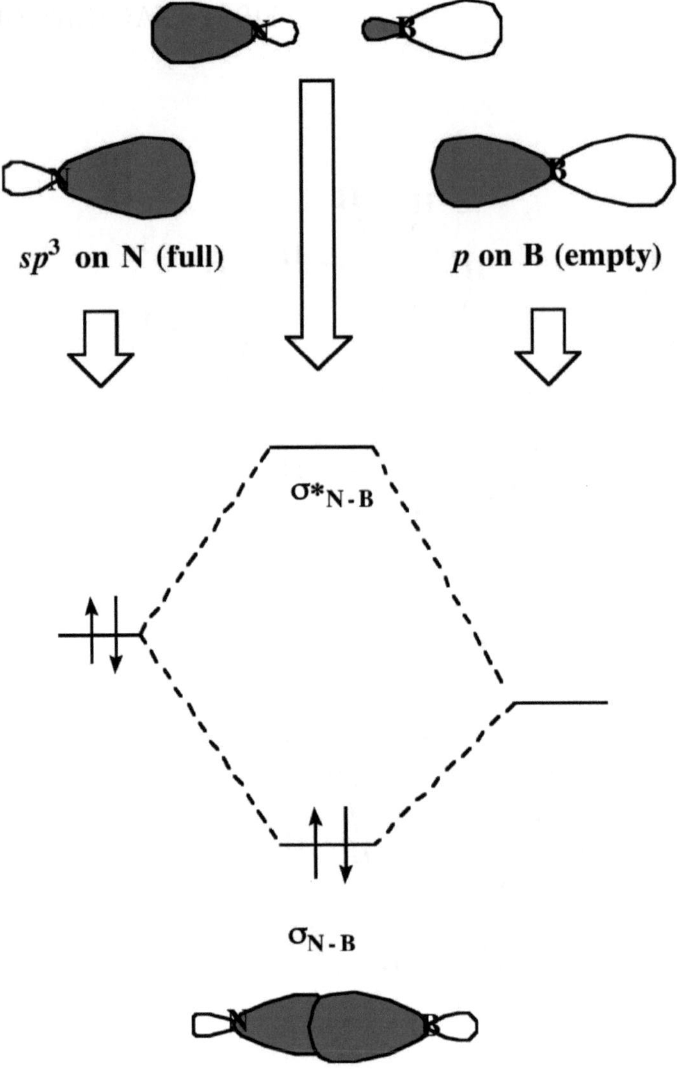

The full (hybrid) atomic orbital from the N interacts with the empty atomic orbital from the B to generate two new molecular orbitals, one that is full (bonding) of lower energy than the initial atomic orbitals, and the other empty (antibonding) of higher energy than either of the initial atomic orbitals. The new molecular orbitals (both bonding and antibonding) have cylindrical symmetry about the internuclear axis and thus are σ molecular orbitals.

3. (20 points)

For benzyl alcohol:

%C and %H -

C 7 x 12.011 = 84.077

H 8 x 1.008 = 8.064

O 1 x 15.999 = 15.999

Molecular weight: 108.140

84.077/108.140 = 77.75% C

8.064/108.140 = 7.46% H

In a 6.00 mg sample, thus,

Wt. C = 0.7775 x 6.00 = 4.66 mg
Wt. H = 0.0746 x 6.00 = 0.45 mg

For CO_2,

C 12.011 x 1 = 12.011
O 15.999 x 2 = 31.998

Molecular weight: 44.009

%C in CO_2 = 12.011/44.009 = 27.29%

For H_2O,

H 2 x 1.008 = 2.016
O 1 x 15.999 = 15.999

Molecular weight: 18.015

%H in H_2O = 2.016/15.999 = 11.19%

For CO_2: wt. = 4.66 mg/0.2729 = 17.08 mg
For H_2O: wt. = 0.45 mg/0.1119 = 4.02 mg

4. (20 points)

You would add water to the mixture of the two compounds. The sodium chloride would dissolve in the water, but the triacontane would not, and would remain as a solid. The separation would be completed by filtering the mixture, the triacontane being retained in the filter and the sodium chloride being in the aqueous filtrate. If it were necessary to isolate the pure sodium chloride, the water would need to be evaporated from the filtrate.

This separation procedure is based on the fact that the solubility of NaCl, an ionic solid, in water is relatively high, the ions being solvated heavily by the water and thus breaking down the solid lattice. However, the solubility of triacontane is quite low in water, both the organization of the solid triacontane and that of the liquid water needing to be broken (requiring energy), with very little energy being released upon interaction of the water and individual triacontane molecules.

5. (15 points)

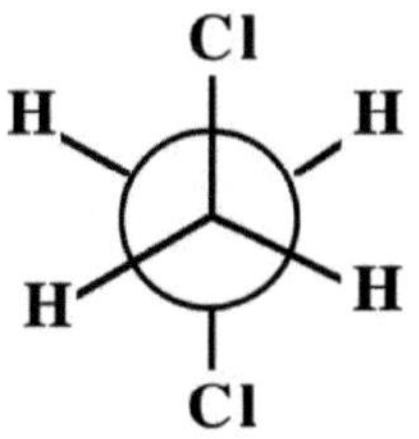

Cl
H H
H H
Cl

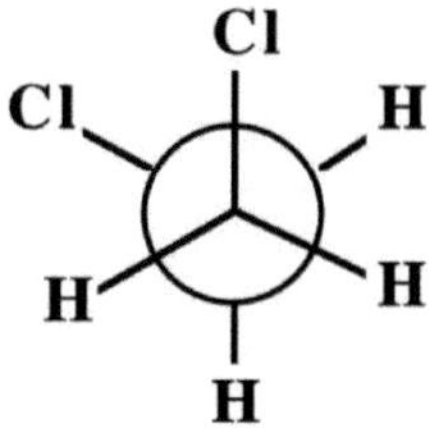

Cl
Cl H
H H
H

We would expect the conformational form shown on the left to be present in the greater amount at any accessible temperature. With the structure shown on the left, the bulky chlorines are as far apart as possible and the bond moments are opposed to each other, whereas with the structure shown on the right the chlorines are relatively close (their associated electrons can repel each other) and the bond moments (vector addition) provide a significant dipole for the form.

6. (15 points)

(a) ethane < ethanol < 1-butanol (lack of hydrogen bonding with ethane, and molecular weight are the factors involved)

(b) pentane < diethyl ether < 1-pentanol (lack of a dipole moment with pentane and hydrogen bonding capabilities are the factors involved)

(c) 3,3-diethylpentane < nonane < 1-bromononane (spherical symmetry with 3,3-diethylpentane and molecular weight are the factors involved)

1. (20 points)

The ratio of C atoms to H atoms can be calculated:

For C:

$87.19/12.011 = 7.259$

For H:

$12.81/1.008 = 12.71$

Then, calculating an empirical formula:

$7.259/12.71 = 1/1.75$ or, empirical formula C_4H_7.

This can not be the molecular formula, but indicates an empirical weight of 55, one-half of the observed molecular weight, so the molecular formula is C_8H_{14}. For a hydrocarbon with no π bonds or rings, the formula C_8H_{16} would apply. Thus, since there are no π bonds, there must be one ring in the compound.

2. (10 points)

(a)

```
                              H
                  H      H    |  H
                   \     |    C'
                    \    |  /   \H
                         C
    H       Br   H       |             H
     \        \ /     H  |       H     /
      C        C       \ C        \ C          H
H  / |  \   /     \   /     \   /     \     /
     |    C         C         C         C
    H     |\       /\        /\         |  \ H
      H - C  H    Br  H     H   H       H
          |\
          H  H
```

(b)

```
                      H
          H     H     |  H
           \    |     C'
            \   |  /    \H
                C
   H   H        |                H
    \ /      H  |       H       /
     C        \ C        \ C           H
H  /   \    /     \    /     \     /
         C          C          C
        /\         /\          |  \ H
      HO  H       H   H        H
```

3. (18 points)

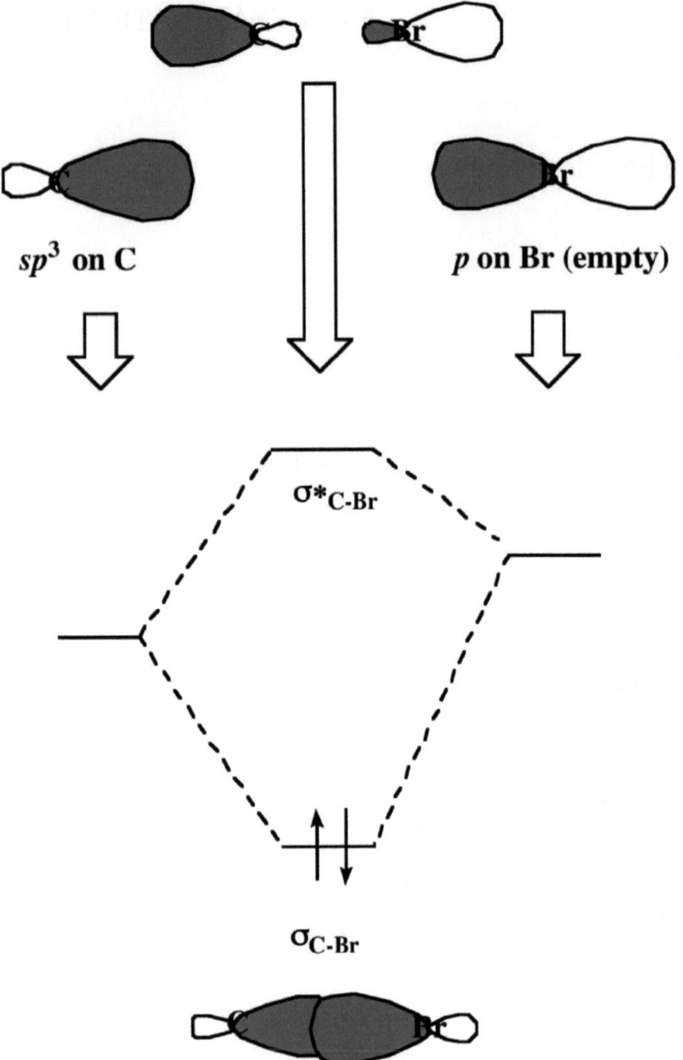

The source of the shared pair of electrons is immaterial; they could both come from either of the original atomic orbitals, or one from each. Both of the resultant molecular orbitals, bonding and anti-bonding, are σ molecular orbitals as they have cylindrical symmetry.

4. (20 points)

Br
H H
H H
Cl

(b)

ClBr
H
H H H

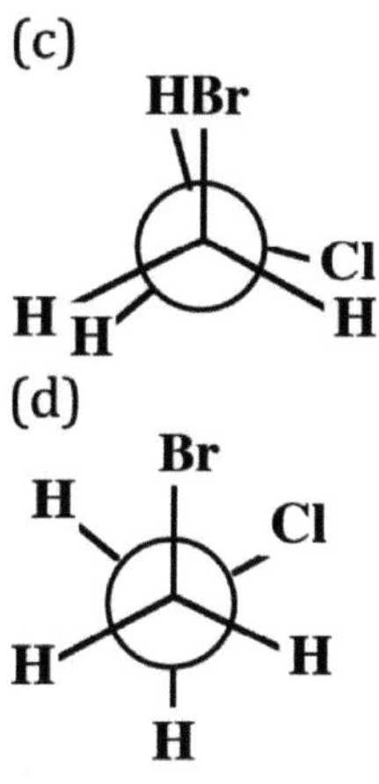

5. (15 points)

sp (b)

(d) sp^2

(e) sp^3

(a) sp^2

sp^3 (c)

6. (17 points)

1-chloro-2-methylpentane

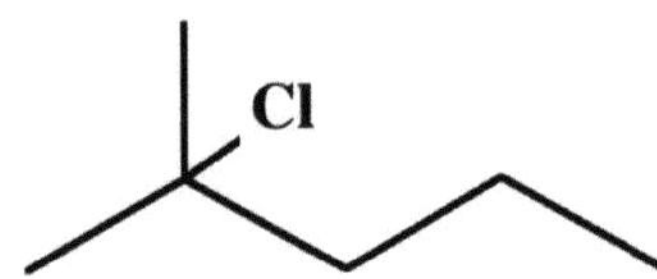

2-chloro-2-methylpentane

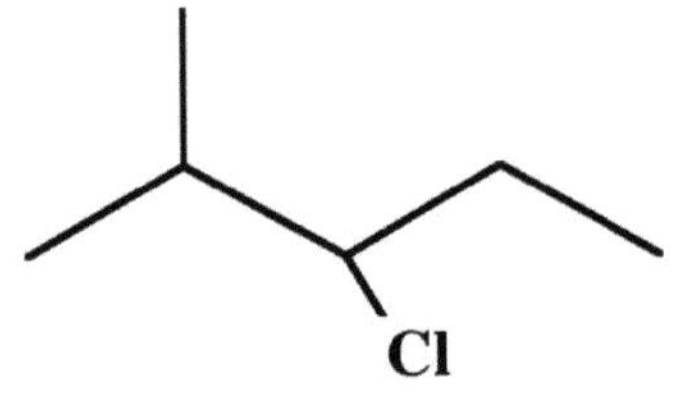

3-chloro-2-methylpentane

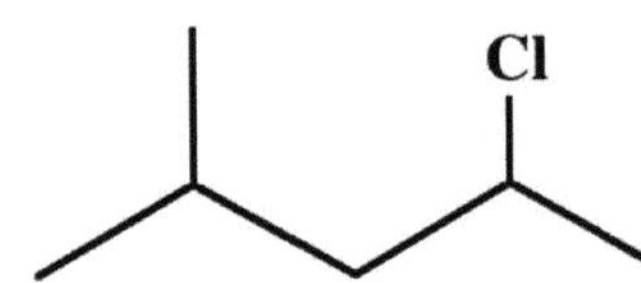

2-chloro-4-methylpentane

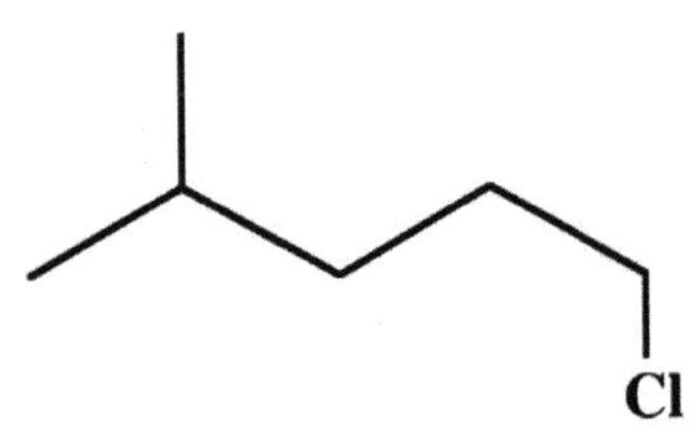

1-chloro-4-methylpentane

1. (20 points)

For a rotation to occur about the C3-C4 bond of *trans*-3-hexene a π bond must be broken. This is a process associated with a high activation energy as the bond must be broken completely before energy is regained by reformation of the bond. However, with the C3-C4 bond of hexane, there is only a σ bond which has cylindrical symmetry and only the non-bonding interactions of electrons in bonds passing by each other in space (through an eclipsed conformation) must be overcome.

2. (20 points)

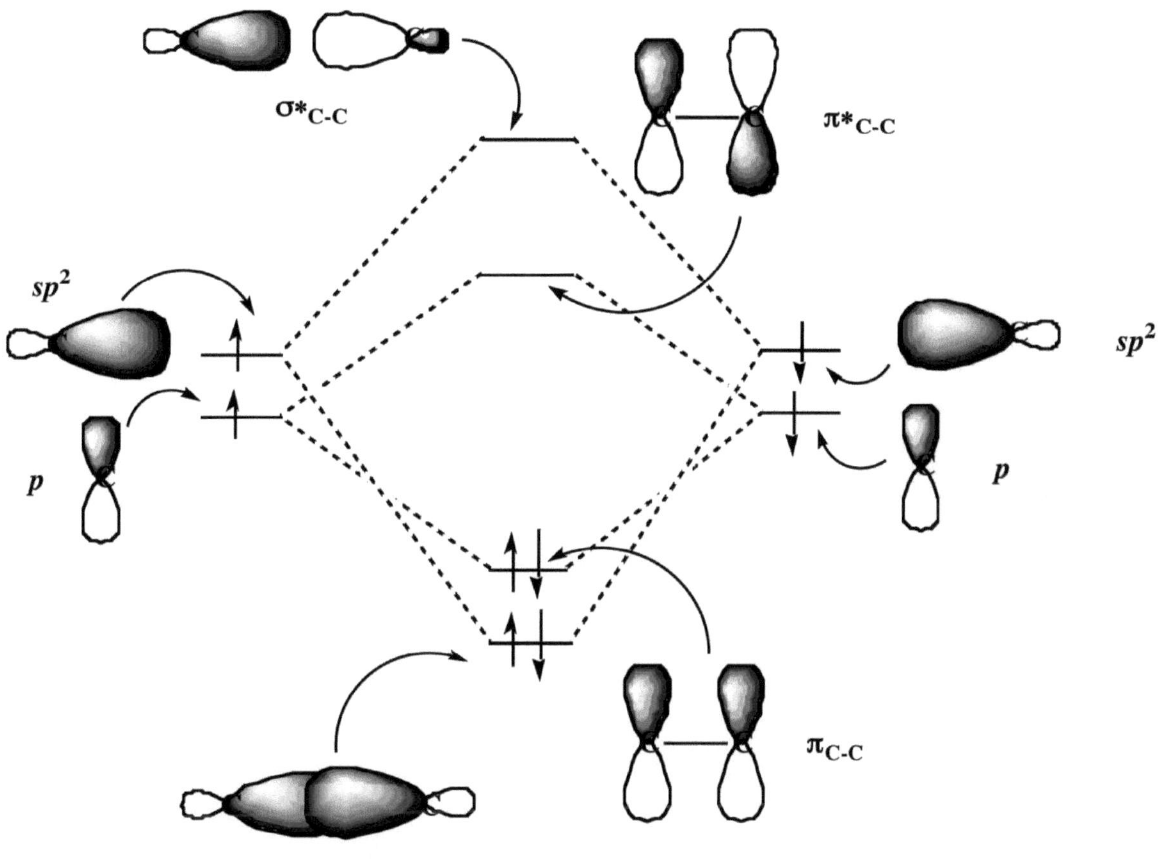

The *p* orbitals generate the π and π* molecular orbitals, while the sp^2 hybrid orbitals generate the σ and σ* molecular orbitals. The π molecular orbitals (bonding and anti-bonding) have a plane of symmetry while the σ molecular orbitals (bonding and anti-bonding) have cylindrical symmetry.

3. (15 points)

The conformation with the hydroxyl groups *anti* relative to each other has no intramolecular stabilizing interaction.

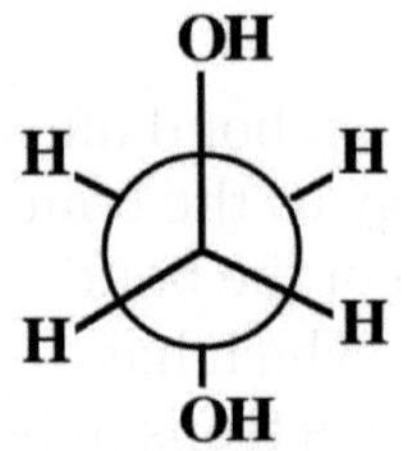

However, with the hydroxyl groups having a 60° dihedral angle between them there is a stabilizing hydrogen bonding interaction that makes it the most stable conformation for the system.

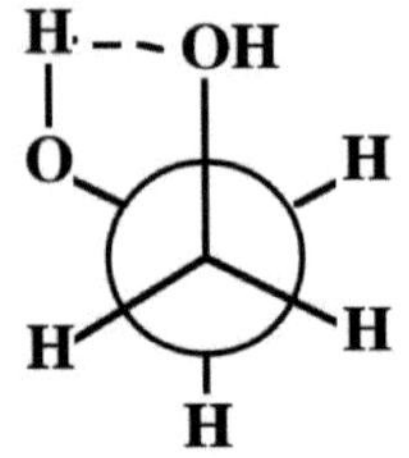

4. (20 points)

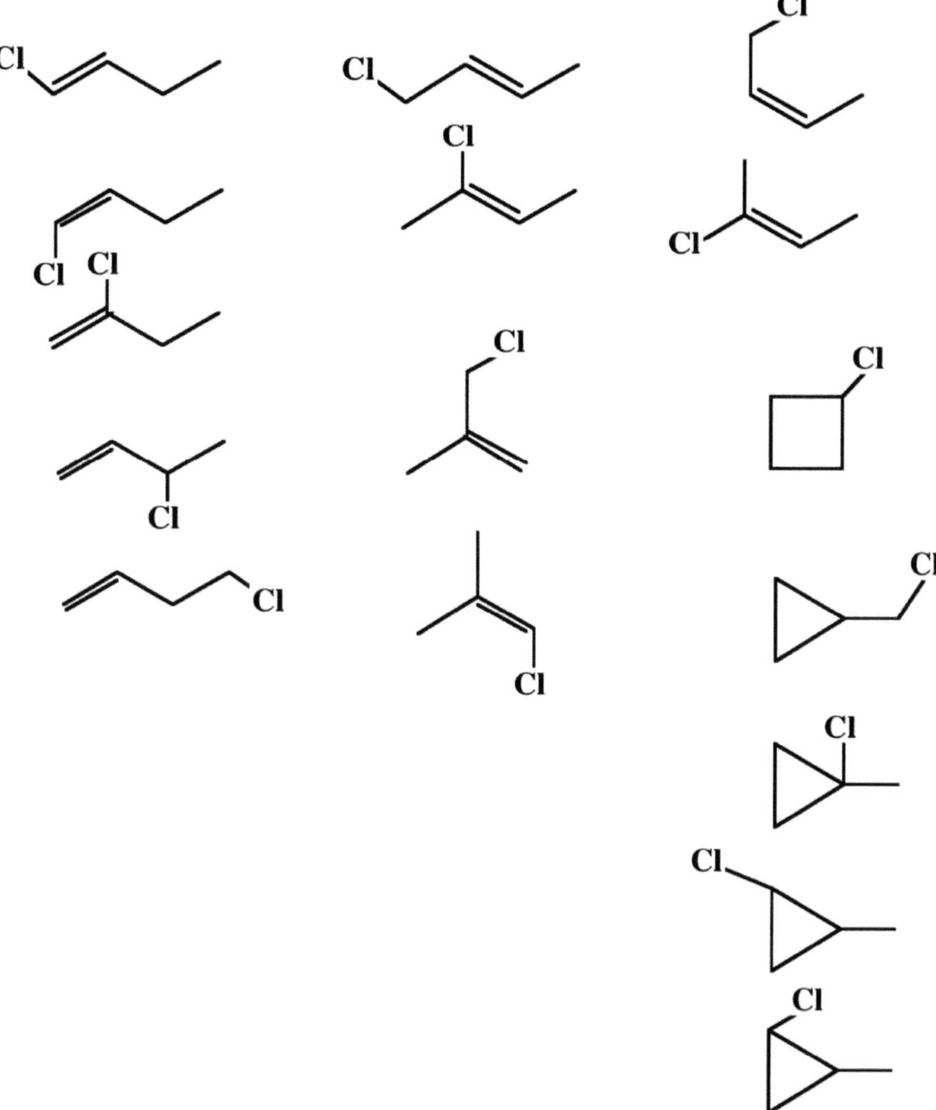

5. (15 points)

(a)

3,3-dimethylpentane - The greater the amount of branching, the greater the stabilization of the hydrocarbon, setting up the situation that less energy is available for release on complete combustion.

(b)

trans-1,2-dimethylcyclopentane - In the *cis* compound the methyl groups are eclipsed rendering it higher in energy than the *trans* compound, providing less energy to be available for the *trans* compound on complete combustion.

(c)

cyclohexane - The presence of the cyclopropyl group with its inherent strain makes it the higher energy of the two compounds, thereby allowing cyclohexane to release less energy on complete combustion.

6. (10 points)

C 6 x 12.011 = 72.066
H 12 x 1.008 = 12.096
Molecular weight = 84.162

%C = 72.066/84.162 = 85.63%
%H = 12.096/84.162 = 14.37%

For both cyclohexane and cycloöctane the *empirical* formula is CH_2, and *empirical* formula is what is directly determined by elemental analysis. The simple combustion analysis would provide the same empirical formula regardless of the sample, be it cyclohexane or cycloöctane.

1. (20 points)

(a) Cyclohexanol $\xrightarrow[H_2SO_4]{CrO_3}$ cyclohexanone

(b) Cyclopentylmethanol ($C_5H_9CH_2OH$) $\xrightarrow[H_2SO_4]{CrO_3}$ $C_5H_9CO_2H$

(c) $(CH_3)_3COH \xrightarrow{H_2SO_4} (CH_3)_2C{=}CH_2$

(d) $(CH_3)_2CH\text{-}CH(Br)\text{-}CH_3 \xrightarrow{K^+\ ^-OC(CH_3)_3}$ 2-methyl-2-butene > 3-methyl-1-butene

(e) $(CH_3)_3C\text{-}OH \xrightarrow[KOH,\ H_2O]{KMnO_4}$ No reaction

2. (20 points)

$$H_3C\text{-}\ddot{O}H + H\text{-}O\text{-}SO_3H \rightleftharpoons H_3C\text{-}\overset{+}{O}H_2 + HSO_4^-$$

$$H_3C\text{-}\ddot{O}H + H_3C\text{-}\overset{+}{O}H_2 \rightleftharpoons H_3C\text{-}\overset{+}{O}(H)\text{-}CH_3 + H_2\ddot{O}$$

$$H_3C\text{-}\overset{+}{O}(H)\text{-}CH_3 + H_2\ddot{O} \rightleftharpoons H_3C\text{-}\ddot{O}\text{-}CH_3 + H_3O^+$$

3. (18 points)

(a) CH_3CH_2, Br, H, OH, H, CH_3 ⇨ Fischer projection: H (top); H_3C—Br; H—OH; CH_2CH_3 (bottom)

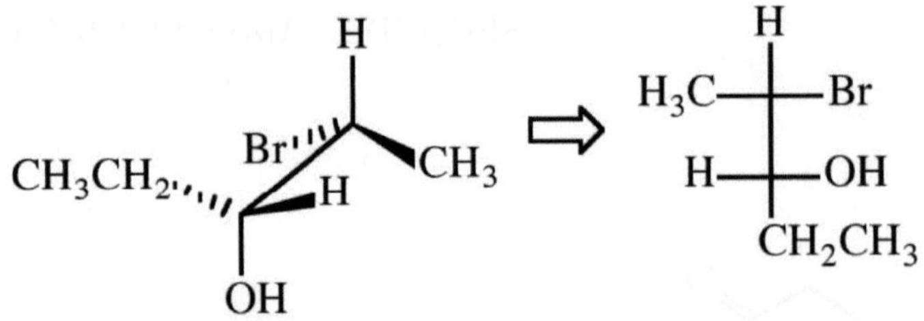

(b) Newman projection: CN, H, Br, OH, CH_3, CH_3 ⇨ Fischer projection: Br (top); H_3C—CN; H—OH; CH_3 (bottom)

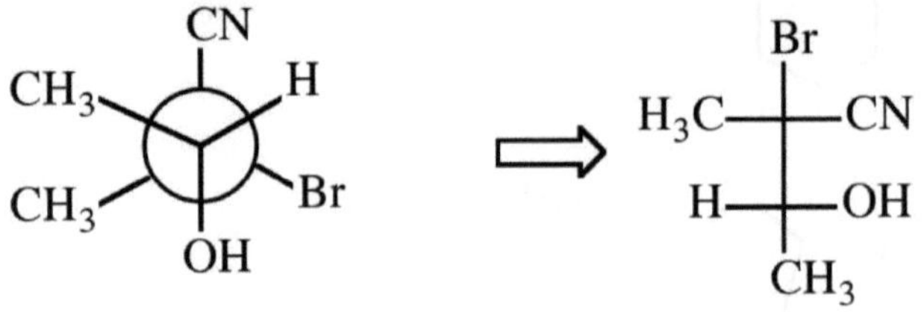

(c) Br, CH_3, H, Cl, H, CH_3 ⇨ Newman projection: H, H_3C, H, H_3C, Br, Cl

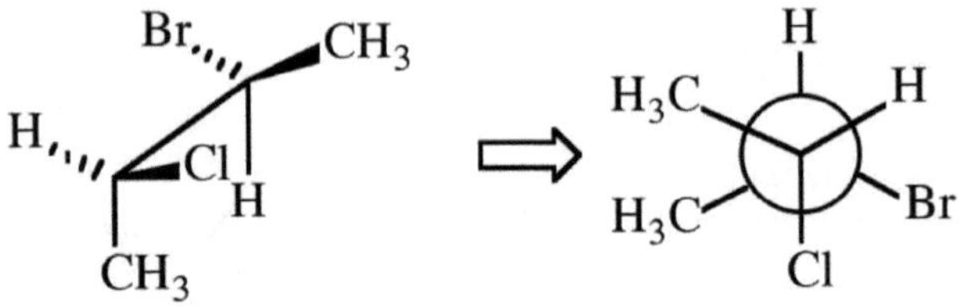

4. (16 points)

(a) I, H, CH_3CH_2, $CH_2CH_2CH_2CH_3$

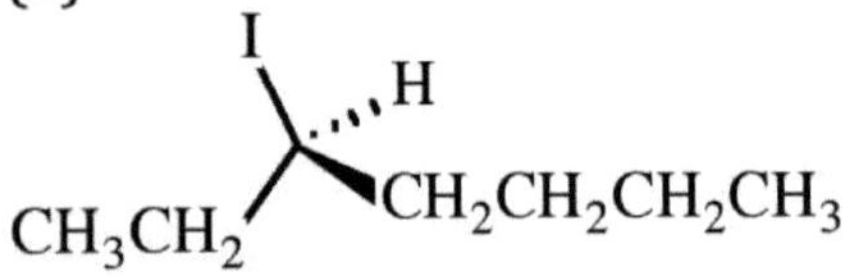

(b) HO, H, $BrCH_2$, CH_2CH_3

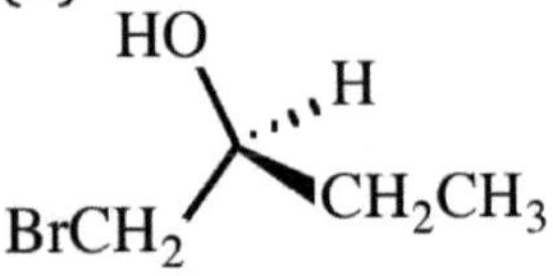

5. (14 points)

(a) $(CH_3)_3CH \xrightarrow[h\nu]{Br_2} (CH_3)_3C\text{-}Br$

(b) $CH_3CH_2CH_2CH_2OH \xrightarrow[\text{pyridine}]{CrO_3} CH_3CH_2CH_2\text{-}\overset{O}{\overset{\|}{C}}\text{-}H$

6. (12 points)

The reaction would immediately generate a 2° carbocation that would rapidly undergo a rearrangement (hydride shift) to generate a 3° carbocation, which would lead to (among other products) the tertiary bromide, 2-bromo-2-methylbutane. In addition to this product, alkenes would also be formed, particularly, 2-methyl-2-butene, 3-methyl-1-butene, and a small amount of 2-methyl-1-butene.

1. (20 points)

(a)

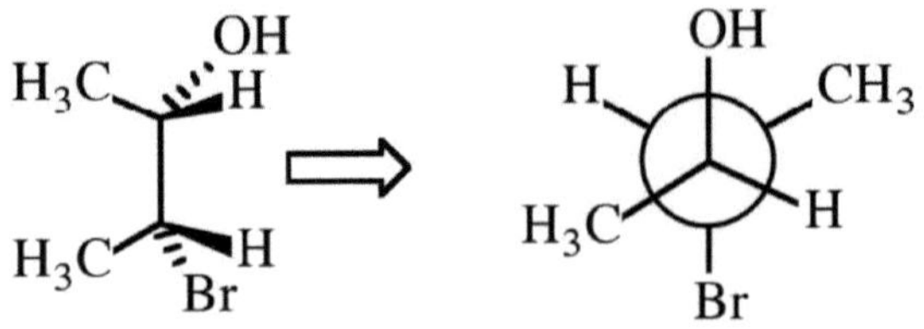

(b)

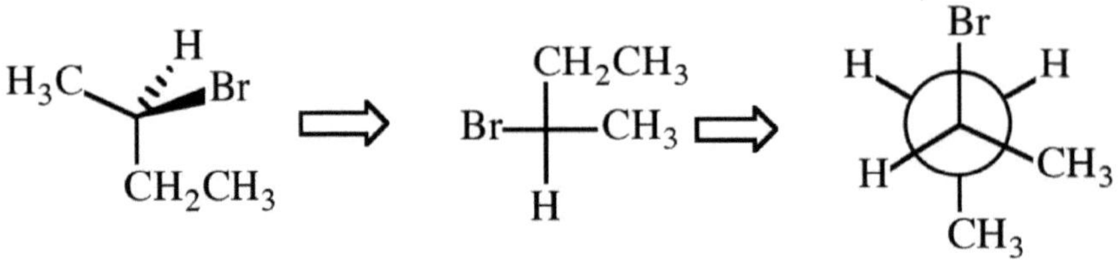

(c)

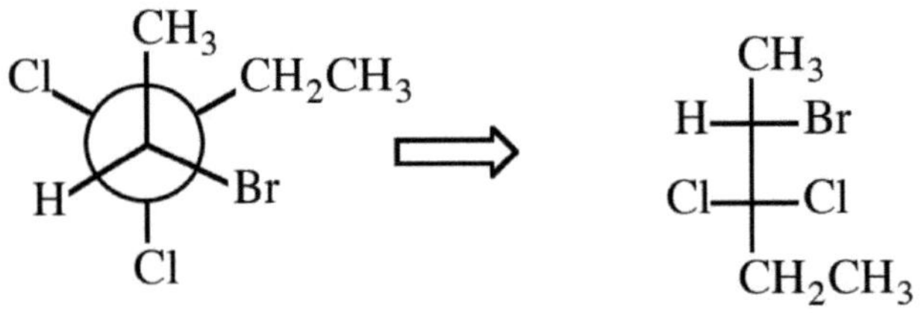

2. (18 points)

(a) activated complex - The structure of the species that exists at the transition state, the highest energy requiring state as reactants are being converted to products.

(b) *meso*-compound - A structure with more than one stereogenic center, but which has an internal plane of symmetry and thus is optically inactive.

(c) diastereoisomers - Stereoisomers that are not enantiomers

3. (20 points)

Carbocations are stabilized by the delocalization of electron density into the electron deficient site. The electron density so delocalized into the electron deficient site originates from the σ (bonding) molecular orbitals of adjacent C-H bonds. These σ (bonding) molecular orbitals of adjacent C-H bonds interact with the empty *p* orbital at the carbocation site as shown. The greater the number of such bonds present (number of adjacent C-H bonds), the greater the electron density that can be delocalized into the empty *p* orbital. Carbocation **A** has nine such C-H bonds whereas carbocation **B** has only two, providing **A** with the greater degree of stabilization.

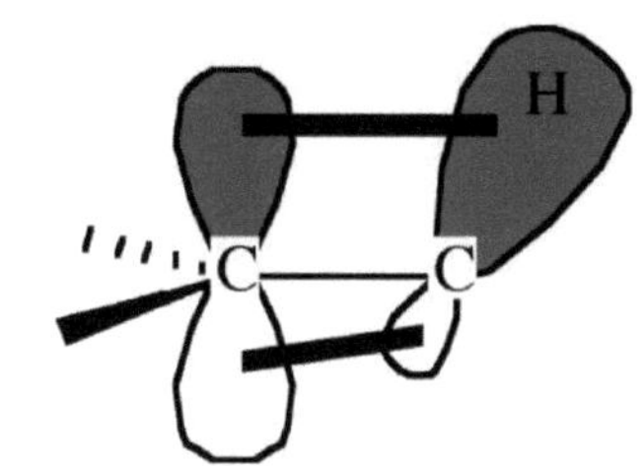

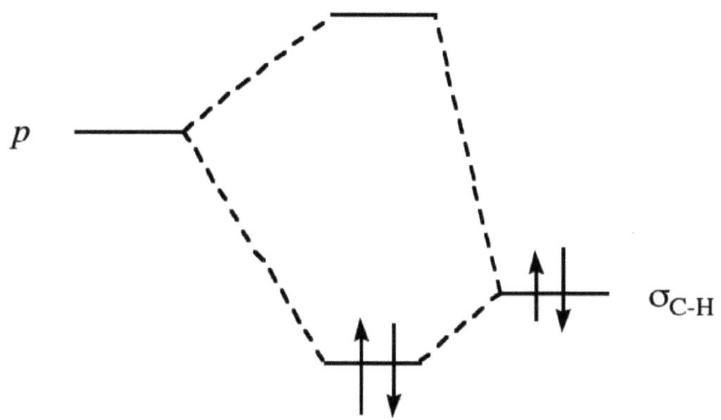

4. (15 points)

(a) $CH_3CH_2CH_2CH_2OH \xrightarrow[\text{pyridine}]{CrO_3} CH_3CH_2CH_2CH{=}O$

(b) OH $\xrightarrow{PCl_3}$ Cl

(c) Br $\xrightarrow[HOC(CH_3)_3]{K^+ \, {}^-OC(CH_3)_3}$

5. (15 points)

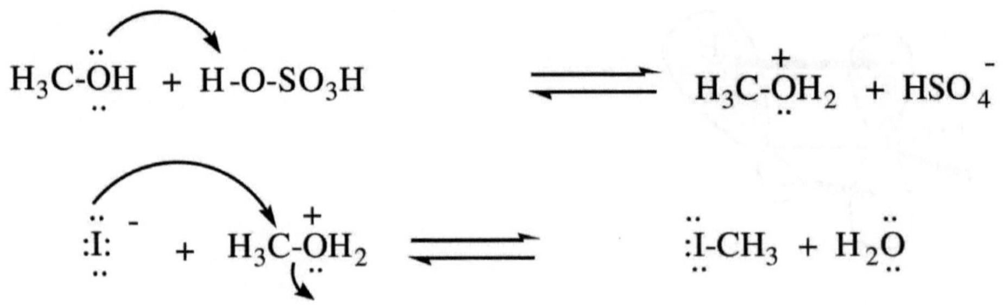

$$H_3C\text{-}\ddot{O}H + H\text{-}O\text{-}SO_3H \rightleftharpoons H_3C\text{-}\overset{+}{O}H_2 + HSO_4^-$$

$$:\ddot{I}:^- + H_3C\text{-}\overset{+}{O}H_2 \rightleftharpoons :\ddot{I}\text{-}CH_3 + H_2\ddot{O}$$

6. (12 points) Give the structure for each of the compounds indicated **C-F**.

C OH

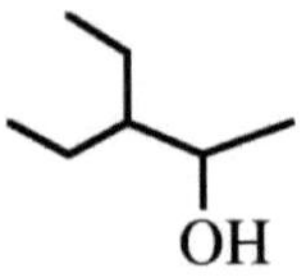

D Br

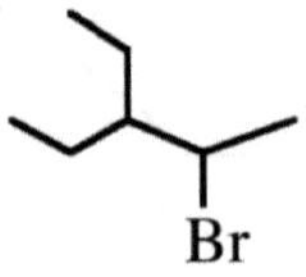

E

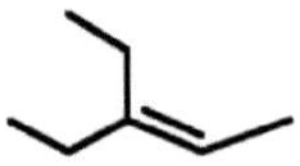

F Br

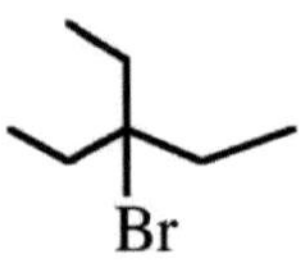

1. (15 points)

(a)

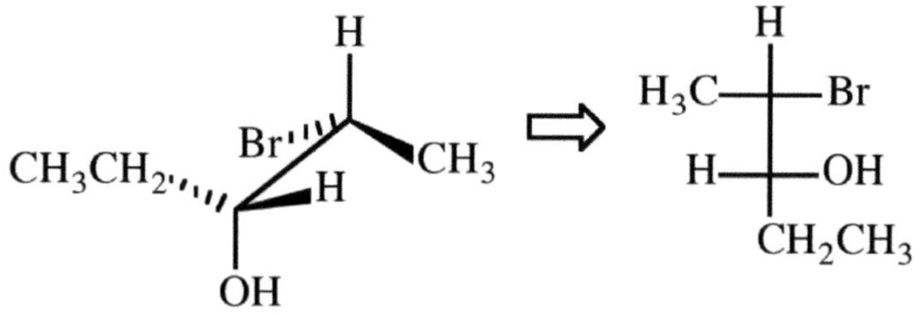

(b)

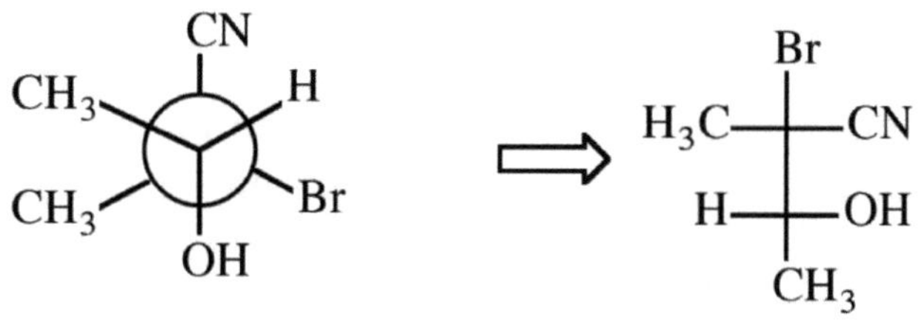

(c)

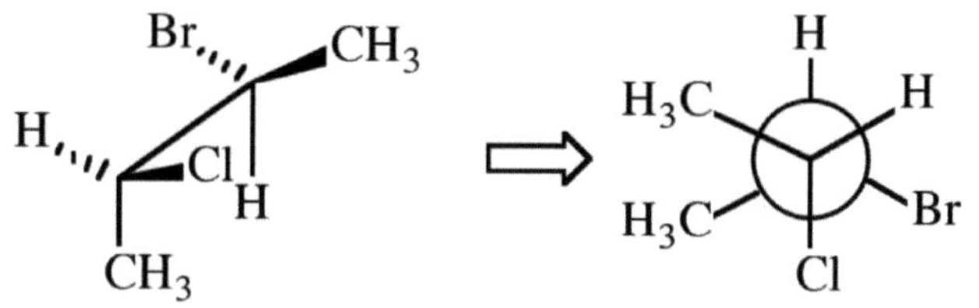

2. (15 points)

(a) E2 Reaction - elimination/bimolecular reaction - it is a β-elimination reaction in which two molecules are involved in the rate determining step

(b) *syn*-hydroxylation - the addition of two -OH units to the same side of an alkene linkage

(c) nucleophile - a Lewis base acting to donate a pair of electrons to an electron deficient site other than a hydrogen

3. (21 points)

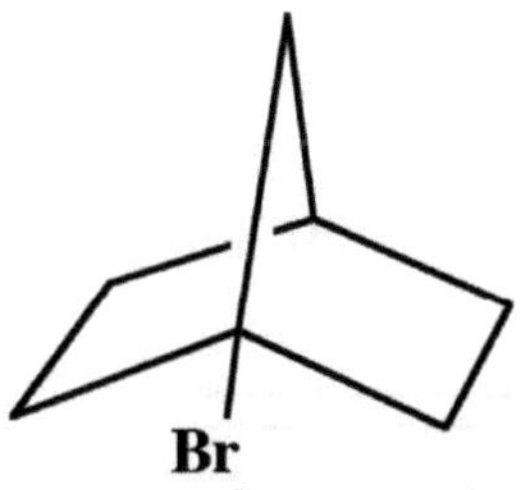

One might consider four possible reaction types:

S_N2 - for this to occur the nucleophile would need to attack the electrophilic carbon from the side opposite the halogen, a feat that is impossible owing to blockage of that side by the remainder of the bicyclic system and that would result in the product having four bonds to carbon all lying in one hemisphere.

S_N1 - this would require a carbocation to be formed - such a carbocation, while 3°, would not have any stabilization by adjacent C-H bonding electrons because planarity about the carbocation site would be required and that feat is impossible with the restrictions of the bicyclic system.

E2 - this would require a *syn*- or *anti*-periplanar array of the C-Br and adjacent C-H molecular orbitals, which is sterically prohibited with the bicyclic system.

E1 - this again would require that a carbocation be formed at the bridgehead site, a species for which stabilization by delocalization of electrons from adjacent C-H bonding molecular orbitals is impossible owing to the geometric restrictions of the bicyclic system keeping the electrophilic carbon site from becoming planar.

4. (12 points)

(a) The reaction would proceed by an E2 route completely (rather than a substitution route as indicated) and give 3-ethyl-2-pentene as the only product.

(b) The reaction would continue oxidizing the aldehyde shown to the carboxylate salt, which is the product that would be isolated.

5. (12 points)

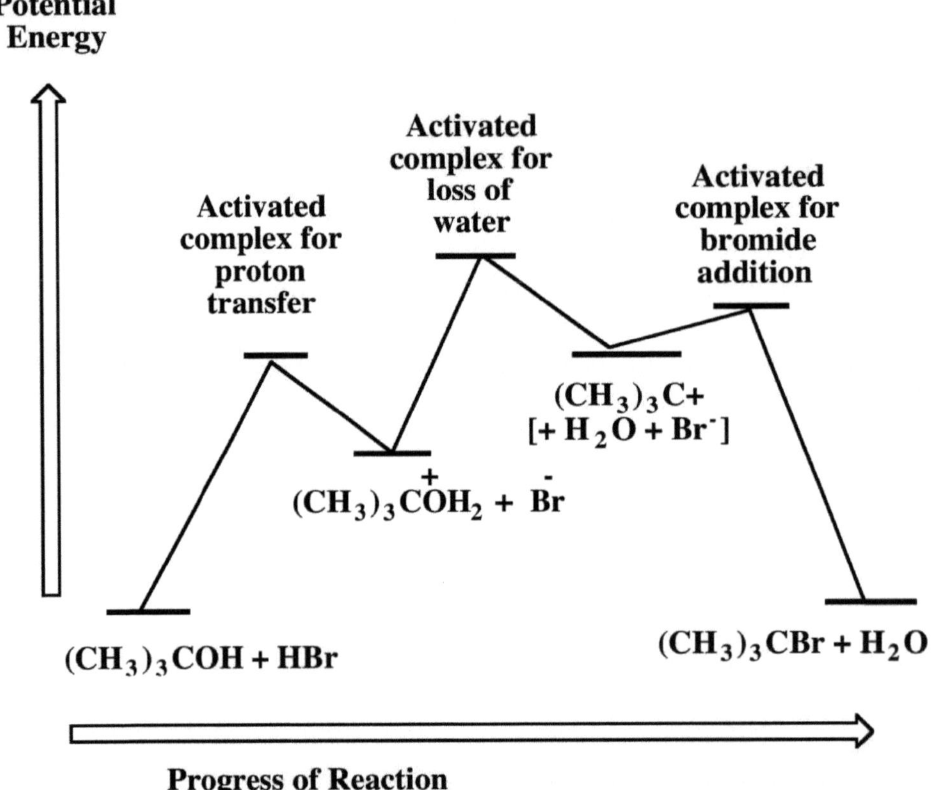

The activated complexes are always higher in energy than either the starting material leading to them or the product generated from them. The highest energy species as an intermediate is the carbocation, higher in energy than the oxonium ion from initial protonation of the alcohol. The activation energy for carbocation leading to final product is relatively low.

6. (25 points)

A

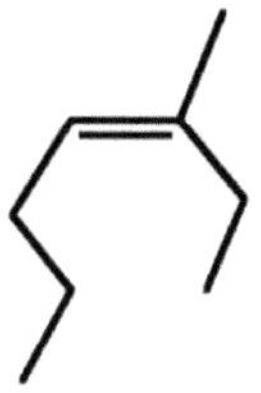

or

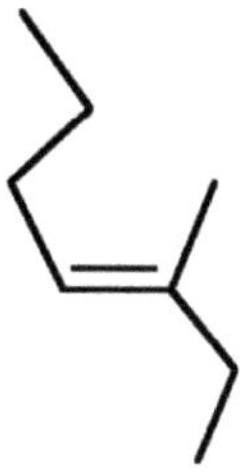

B

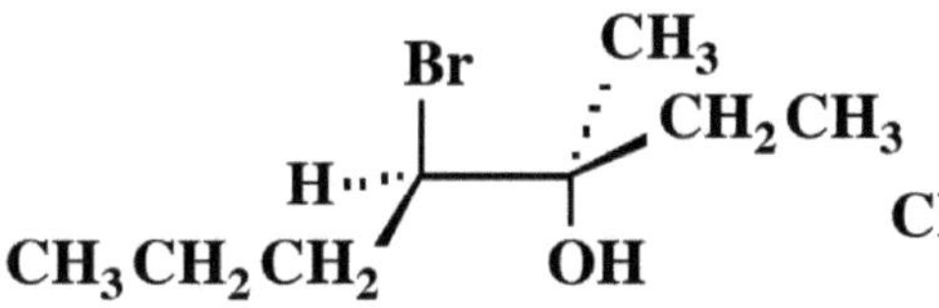

Br CH_3 CH_2CH_3 H $CH_3CH_2CH_2$ OH

(racemic)

or

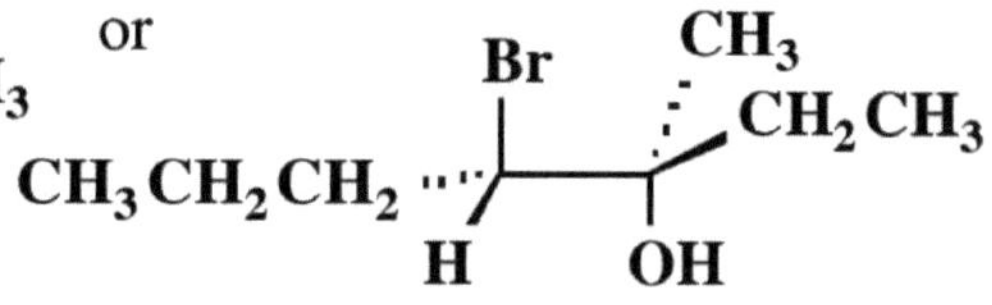

Br CH_3 CH_2CH_3 $CH_3CH_2CH_2$ H OH

(racemic)

C

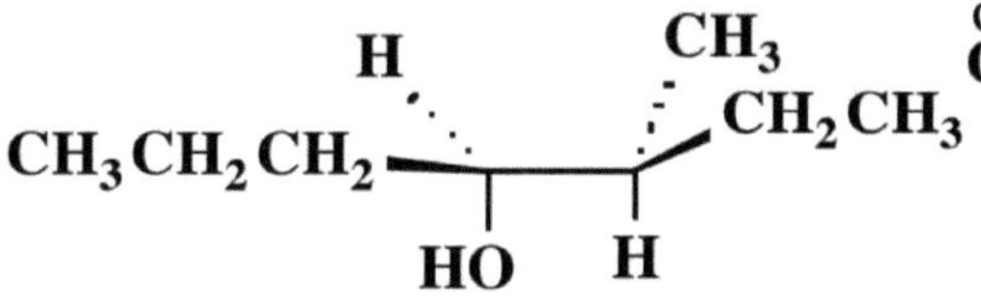

H CH_3 CH_2CH_3 $CH_3CH_2CH_2$ HO H

(racemic)

or

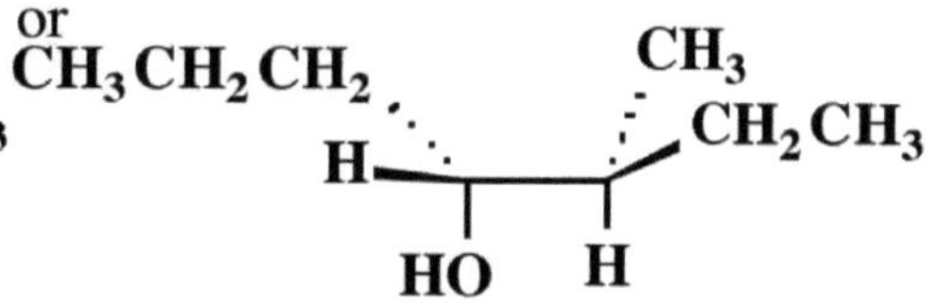

$CH_3CH_2CH_2$ CH_3 CH_2CH_3 H HO H

(racemic)

D

OH

(racemic)

E

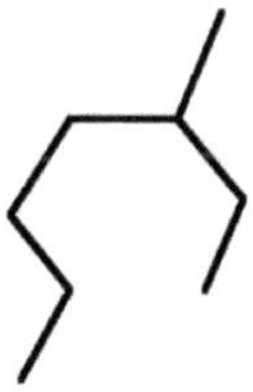

(racemic)

The geometry of the alkene (**A**) can not be determined by the ozonolysis reaction - as a result, there are two sets of "correct" answers.

ORGANIC CHEMISTRY I THIRD EXAMINATION (I)
ANSWERS

1. (25 points)

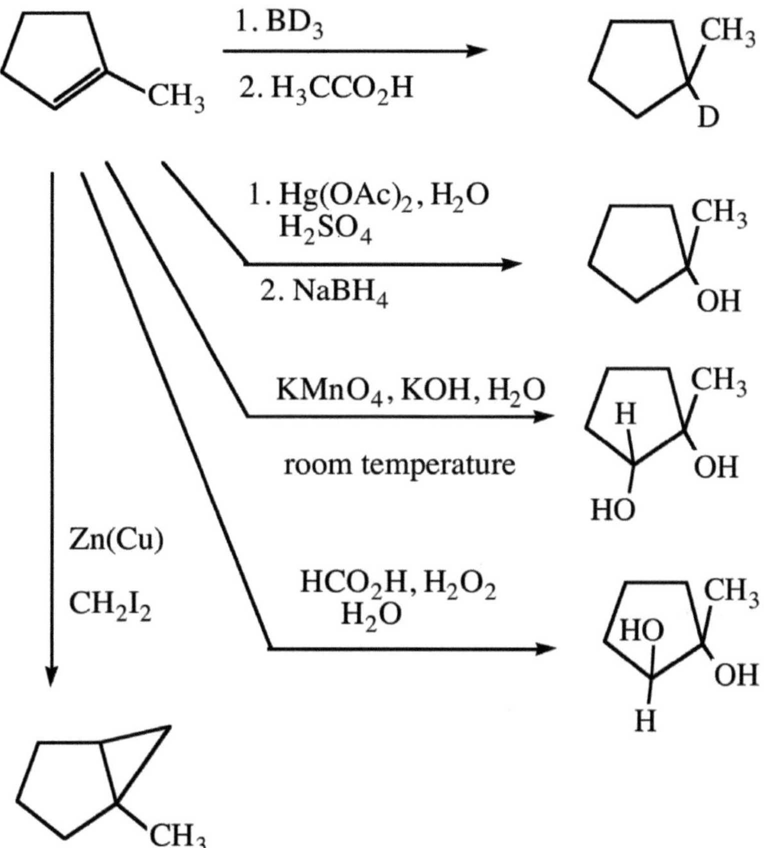

2. (12 points)

CH3 CH3 CH3 H3C CH3 CH3

less stable more stable

3. (25 points)

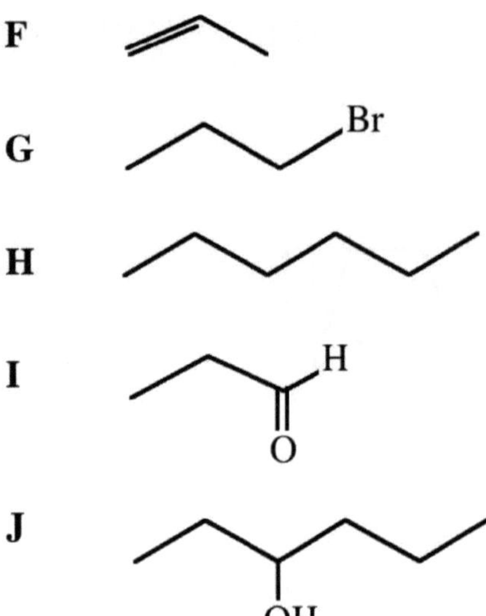

4. (25 points)

(a) (*R*)-3-chlorohexane

(b) 3-ethyl-2-pentene

(c) 3-ethyl-3-pentanol (or ethyl 3-ethyl-3-pentyl ether, depending on the amount of ethanol used)

(d)

H H

CN

H_3C

(e) ethyl 1-hexyl ether

5. (13 points)

$$\mathbf{A} + \mathbf{B} \xrightarrow{\text{slow}} \mathbf{C}$$

$$\mathbf{C} + \mathbf{A} \xrightarrow{\text{rapid}} \mathbf{D} + \mathbf{E}$$

1. (25 points)

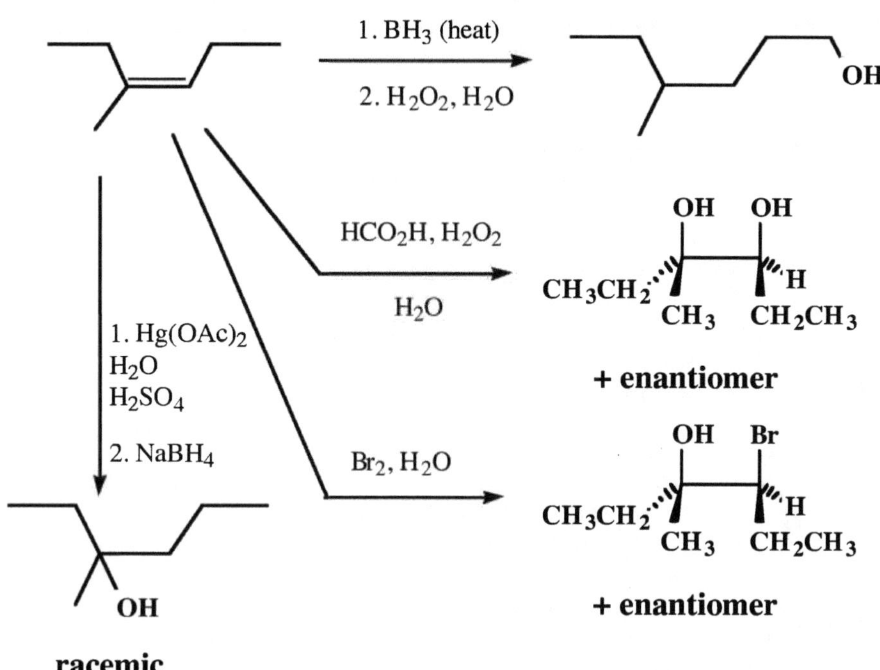

2. (20 points)

No β-elimination can occur because:

(1) For an E1 reaction to occur there must be generated a carbocation by loss of the bromide anion from carbon. Such a resultant 2° carbocation requires stabilization by filled C-H molecular orbitals donating electrons from adjacent C-H sites. Such adjacent C-H bonding mole cular orbitals are not in position to overlap with the empty *p* orbital of the presumed carbocation site.

(2) For an E2 reaction to occur there must be an adjacent C-H linkage that can be either *syn*- or *anti*-periplanar relative to the C-Br linkage. The adjacent C-H linkages have improper dihedral angles rather than 0° (*syn*) or 180° (*anti*) with regard to the C-Br linkage.

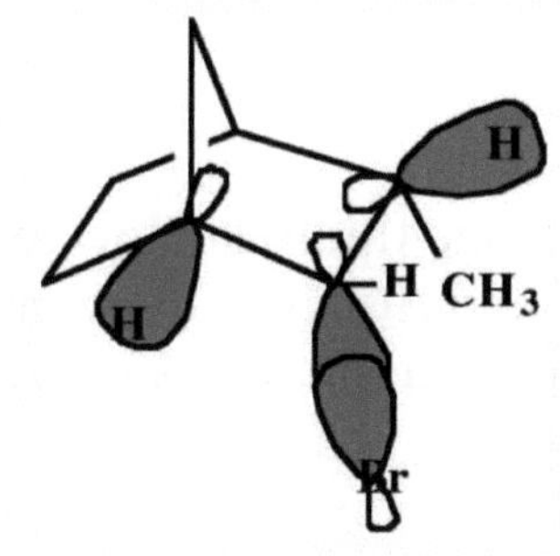

As a result, no π bond formation can occur to support the simultaneous breakage of the C-H and C-Br bonds.

3. (30 points) Give the structure for each of the compounds indicated **E-J**.

E

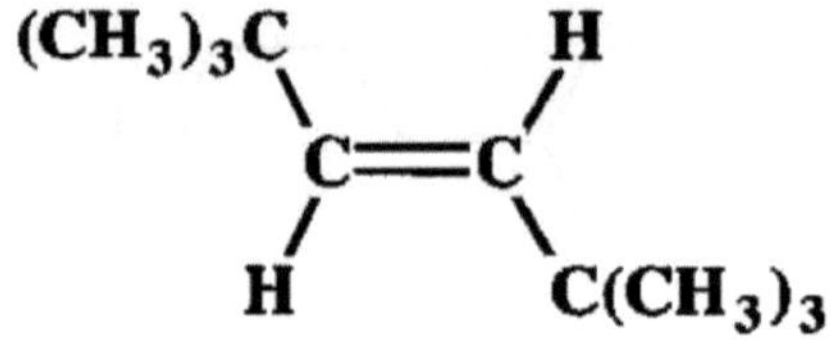

$$\begin{array}{ccc} (CH_3)_3C & & H \\ & C{=}C & \\ H & & C(CH_3)_3 \end{array}$$

F $(CH_3)_3CO_2H$

G

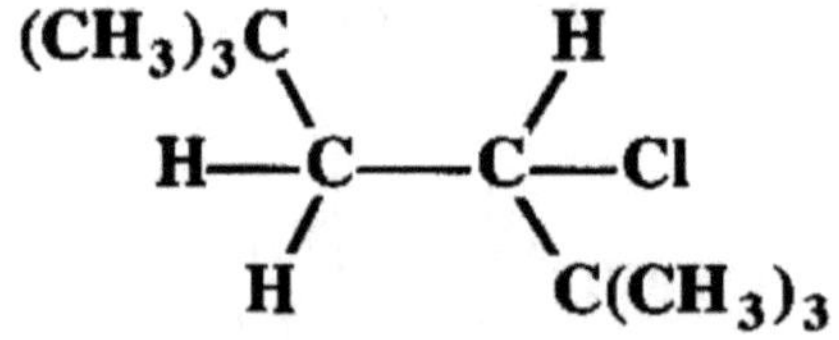

$$\begin{array}{ccc} (CH_3)_3C & & H \\ H{-} & C{-}C & {-}Cl \\ H & & C(CH_3)_3 \end{array}$$

H

$$\begin{array}{ccc} (CH_3)_3C & & C(CH_3)_3 \\ & C{=}C & \\ H & & H \end{array}$$

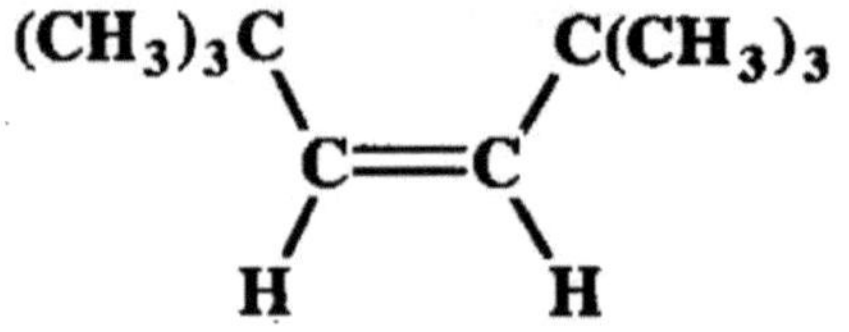

I

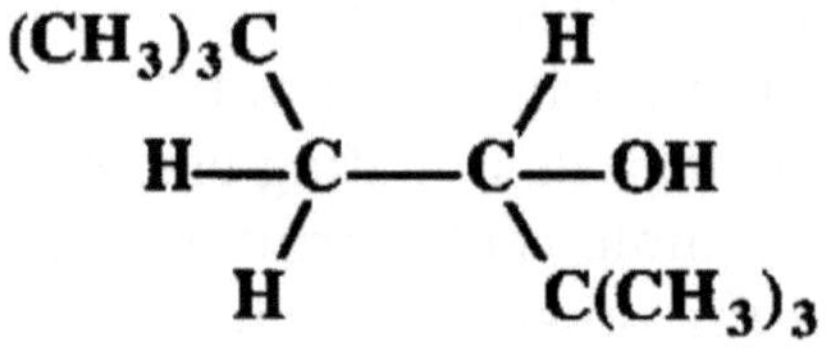

$$\begin{array}{ccc} (CH_3)_3C & & H \\ H{-} & C{-}C & {-}OH \\ H & & C(CH_3)_3 \end{array}$$

J

$$\begin{array}{cccc} & OH & OH & \\ (CH_3)_3C & C & C & H \\ & H & & C(CH_3)_3 \end{array}$$

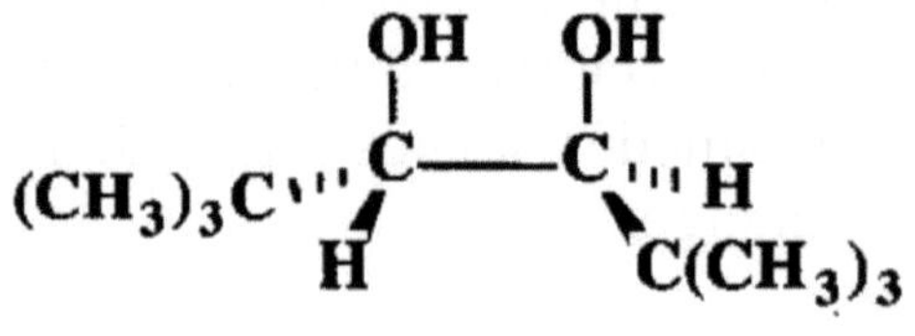

4. (25 points)

The reaction proceeds in three stages. In the first stage thionyl chloride reacts with the alcohol to produce the chlorosulfite ester and HCl. In the second stage the intermediate ester is attacked in an S_N2 manner from the backside by the solvent dioxane (present in very high concentration) to generate a new intermediate oxonium ion. In the final step this intermediate oxonium ion undergoes a second S_N2 reaction with chloride ion (produced from the displaced group in the second stage) to give the observed product. There are two inversion reactions at the original carbinol carbon providing overall retention of configuration at that carbon.

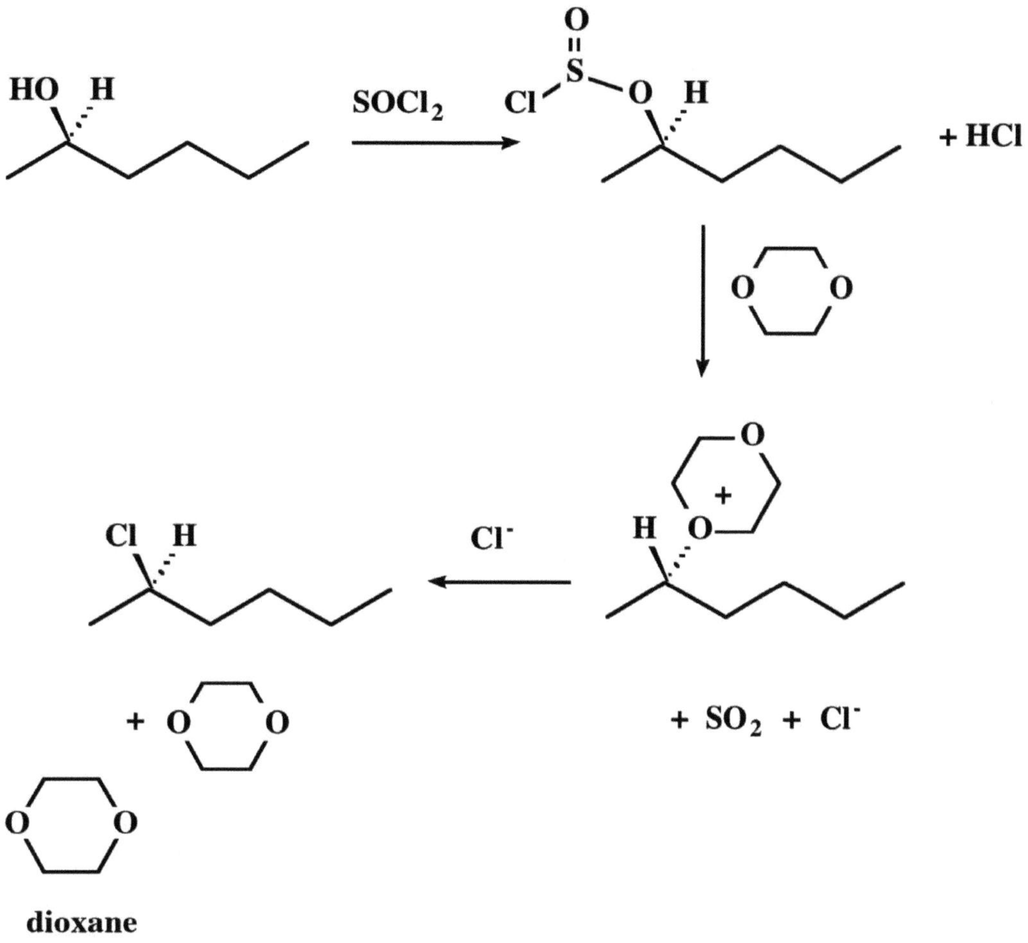

HO H

$SOCl_2$

Cl S O H

O

+ HCl

O O

O

+

H O

Cl H

Cl^-

+ O O

+ SO_2 + Cl^-

O O

dioxane

ORGANIC CHEMISTRY I THIRD EXAMINATION (III)

ANSWERS

1. (25 points)

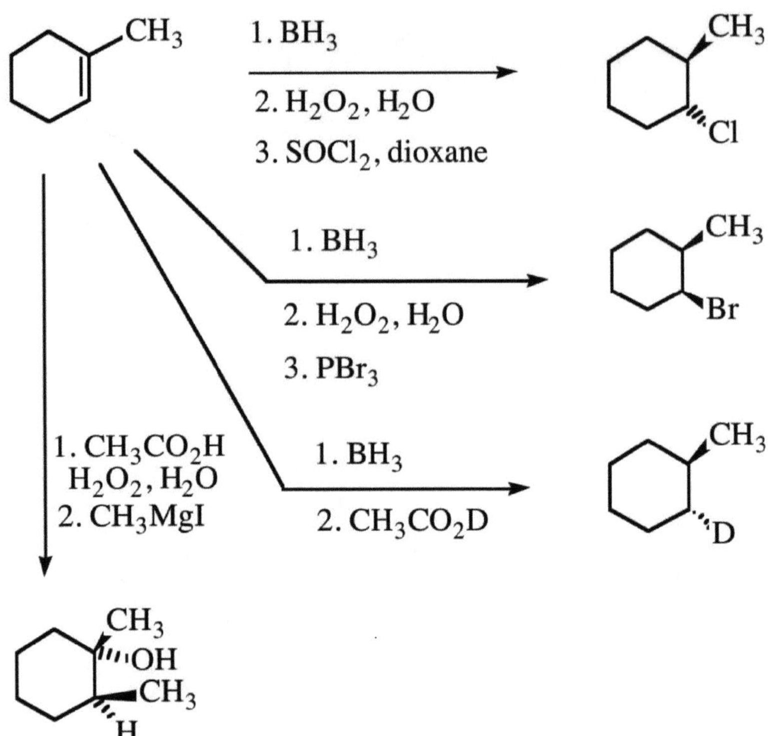

2. (25 points)

(a)

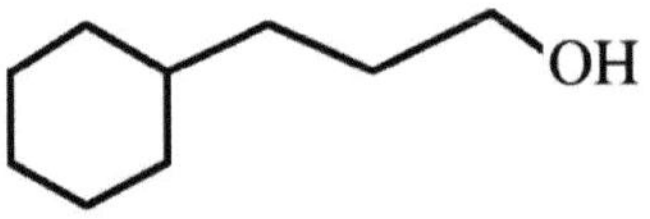

(b)

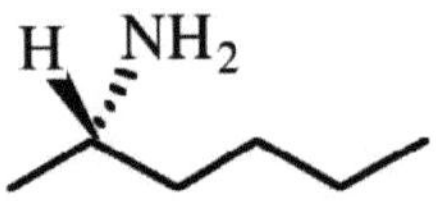

(c)

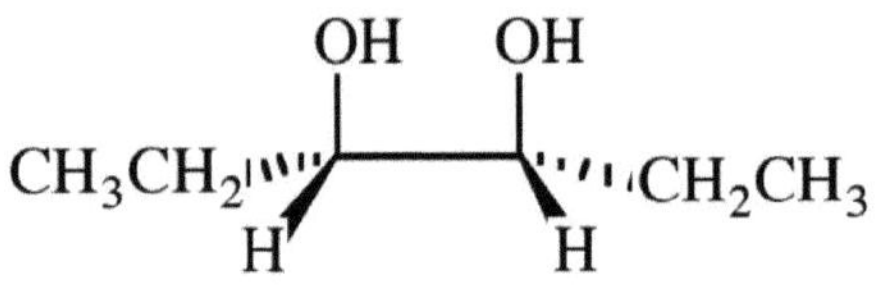

(d) $CH_3CH_2CO_2H$

(e)

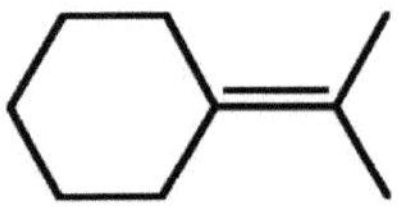

3. (25 points)

E	1. BH_3 2. H_2O_2, KOH, H_2O
F	HBr
G	$NaBH_4$ *or* $LiAlH_4$ *or* H_2, PtO_2
H	1. $CH_3CH_2CH_2MgBr$ 2. H_2O
I	$CrO_3/H_2SO_4/H_2O$ *or* $KMnO_4/KOH/H_2O$
J	1. Mg/diethyl ether 2. D_2O

4. (25 points)

(a) 3-ethyl-3-pentanol and 1-heptanol - add each to CrO_3/aq. sulfuric acid; the original orange color will change to green with 1-heptanol but there will be no color change with 3-ethyl-3-pentanol (alternative: add to $KMnO_4/KOH/H_2O$; the purple color of the solution will be dispersed by oxidation of 1-heptanol, but 3-ethyl-3-pentanol will not change the color since it's not oxidized).

(b) dibutyl ether and nonane - add each to aqueous acid; the dibutyl ether will dissolve in the acid but the nonane will not dissolve.

(c) 3-pentanone and pentanal - add each to CrO_3/aq. sulfuric acid; the original orange color will change to green with pentanal but there will be no color change with 3-pentanone (alternative: add to $KMnO_4/KOH/H_2O$; the purple color of the solution will be dispersed by oxidation of pentanal, but 3-pentanone will not change the color since it's not oxidized).

(d)

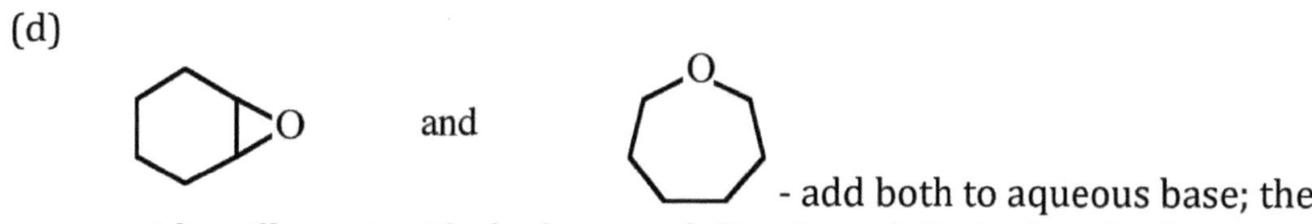

- add both to aqueous base; the cyclohexene oxide will react with the base and dissolve while the hexahydrooxepin (seven-membered ring ether ring system) will not react with base and will not dissolve.

1. (20 points)

The bond may be considered as being generated using an sp^3 hybrid orbital from carbon and a *p* orbital from iodine. We mathematically mix the two functions to generate two new two-centered (molecular) orbitals, one lower in energy than either of the original orbitals (a bonding molecular orbital) and one higher in energy than the original orbitals (an antibonding molecular orbital).

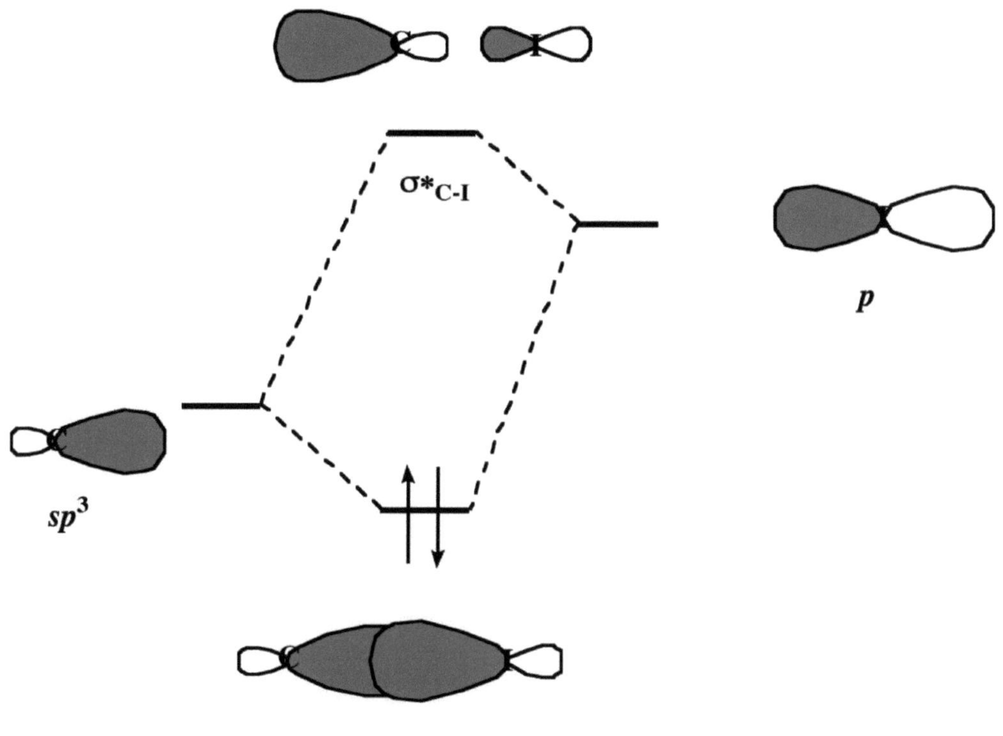

It matters not what the source is of the two electrons in the bonding molecular orbital - we may consider both from either of the two original atomic orbitals, or one from each of them.

2. (26 points)

Water would be added to the mixture of ethanol and hexane. The ethanol would dissolve in the water as it is miscible in water in all proportions, while the hexane, fundamentally insoluble in water, would float on top of the ethanol/water solution. The layers would be separated using a separatory funnel. Some of the ethanol would remain in the hexane. Thus, the above noted procedure would need to be repeated several times to remove virtually all of the ethanol from the hexane. The thus separated hexane after these extractions would be dried over anhydrous magnesium sulfate, separated from the drying agent by filtration, and finally purified by distillation. Isolation of the ethanol from the water could be done by direct distillation. A 95% ethanol/5% water azeotrope would distill first. The remaining water could be removed by adding benzene and distilling first the ternary azeotrope of ethanol/benzene/water after which pure ethanol could be distilled. Water is used here as ethanol can dissolve in water owing to its polar nature and its capabilities for hydrogen bonding whereas hexane is not able to undergo such interactions with water to allow it to dissolve.

3. (15 points)

(a)

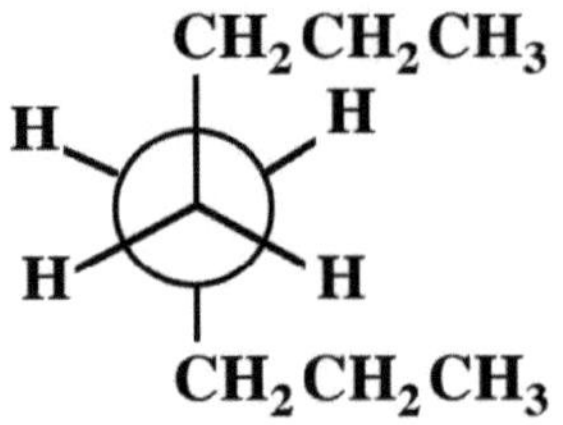

$CH_2CH_2CH_3$ (front), H, H; H, H, $CH_2CH_2CH_3$ (back)

(b)

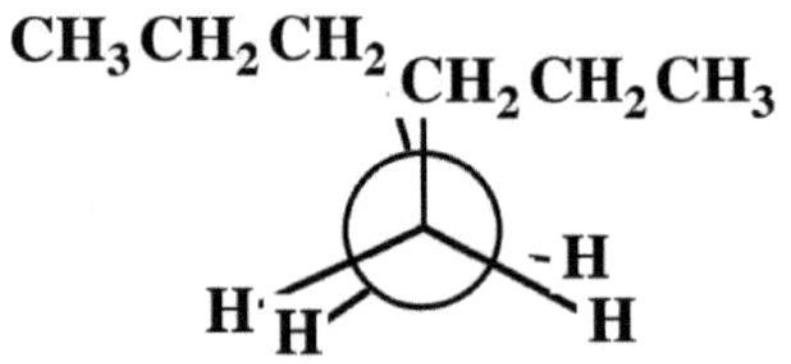

$CH_3CH_2CH_2$, $CH_2CH_2CH_3$; H, H, H, H

(c)

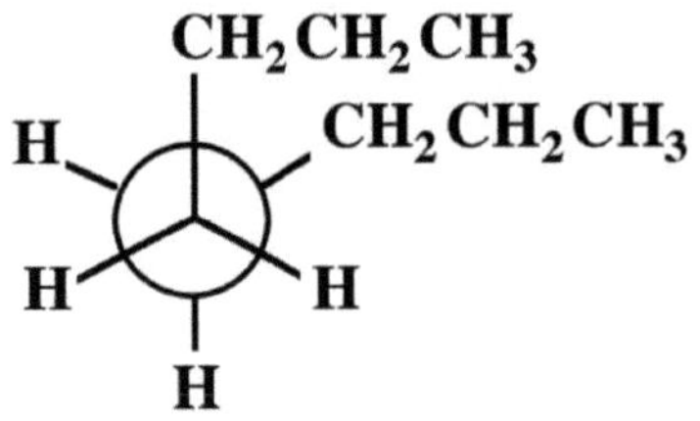

$CH_2CH_2CH_3$, $CH_2CH_2CH_3$; H, H, H, H

4. (22 points)

(a)

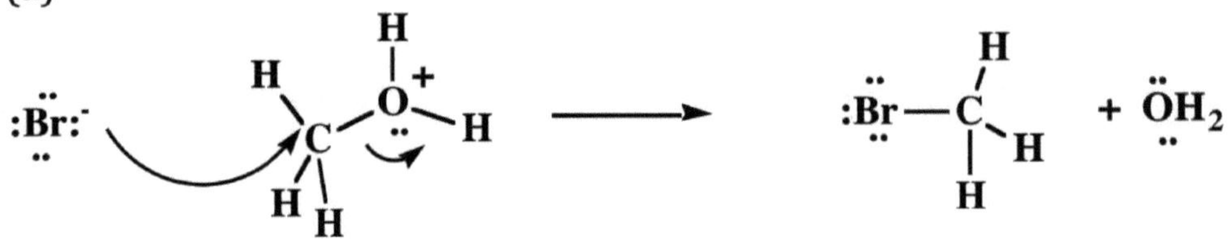

$:\ddot{Br}:^-$ + $CH_3-\overset{+}{O}H_2$ ⟶ $:\ddot{Br}-CH_3$ + $\ddot{O}H_2$

(b)

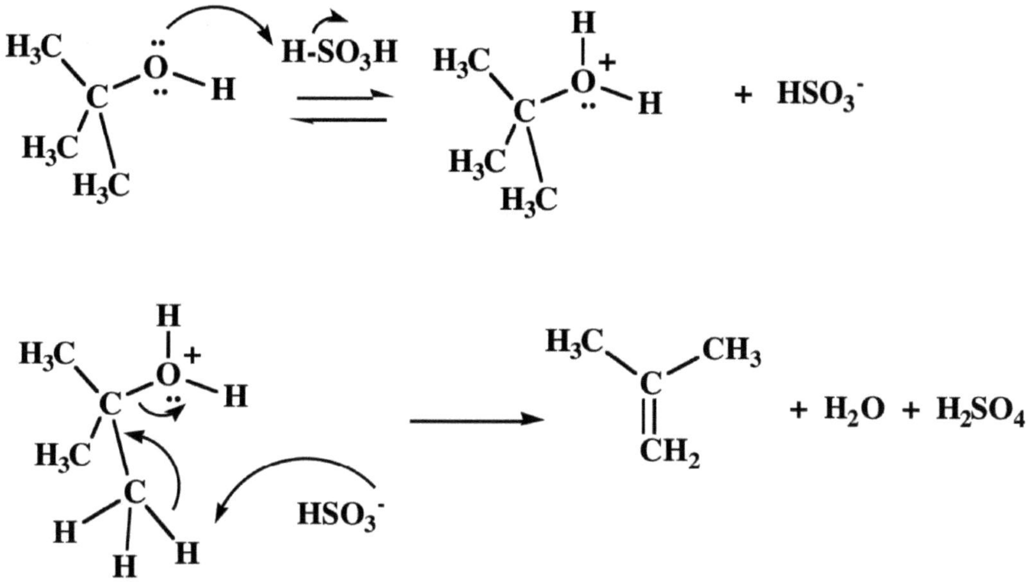

5. (28 points)

(a)

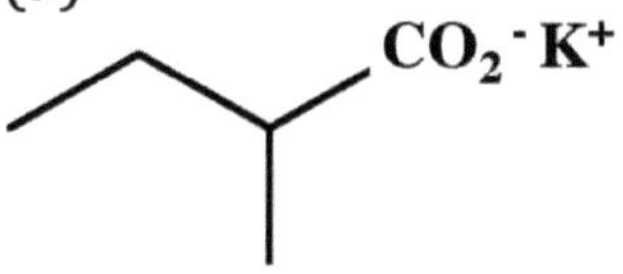

$CO_2^- K^+$

(b)

No Reaction

(c)

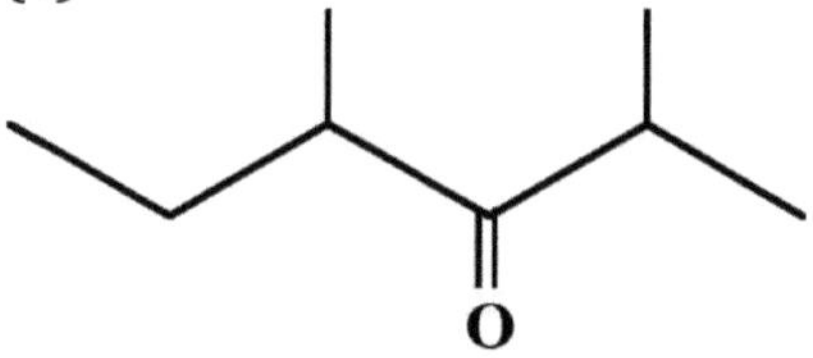

O

(d)

$CO_2^- K^+$

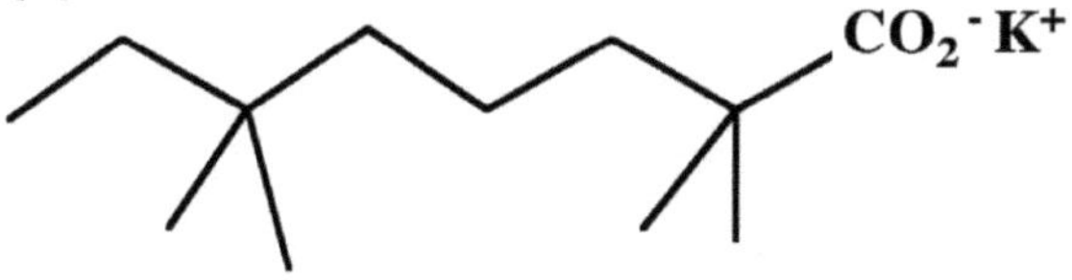

6. (28 points)

(a)

Cl
H CH$_2$CH$_3$
H$_3$C H
Br

⇩

Cl
H$_3$C—H
CH$_3$CH$_2$—H
Br

(b)

CH$_3$
H Cl
H Br
CH$_3$

⇩

Cl
H$_3$C—H
H—CH$_3$
Br

7. (25 points)
Apparently the student is thinking that the 1-hexyloxonium ion would form and undergo either loss of water to give the 1-hexylcation or be attacked by a Lewis base present (the conjugate base of the starting acid?) to remove a hydrogen from the 2-position. It is unlikely that the 1-hexyl cation would form to any appreciable extent and the oxonium ion would be more likely to undergo attack by another molecule of 1-hexanol leading to dihexyl ether. A better route would be first to convert the 1-hexanol to 1-bromohexane with phosphorus tribromide followed by E2 type elimination using potassium *tert*-butoxide.

8. (24 points)
(a)
1. PBr_3 2. $KOC(CH_3)_3$ 3. $KMnO_4$, KOH, H_2O 4. aq. acid
(b)
1. $KOC(CH_3)_3$ 2. BH_3, heat 3. H_2O_2, H_2O, KOH
(c)
1. PBr_3 2. NaCN
(d)
1. NaI, acetone 2. NaCN

9. (18 points)
(a)

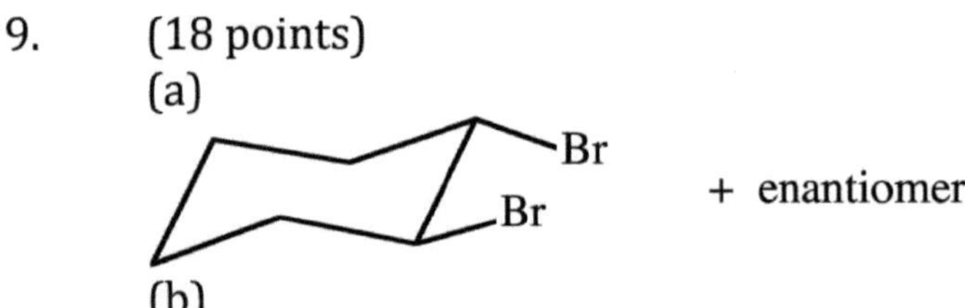

(b)

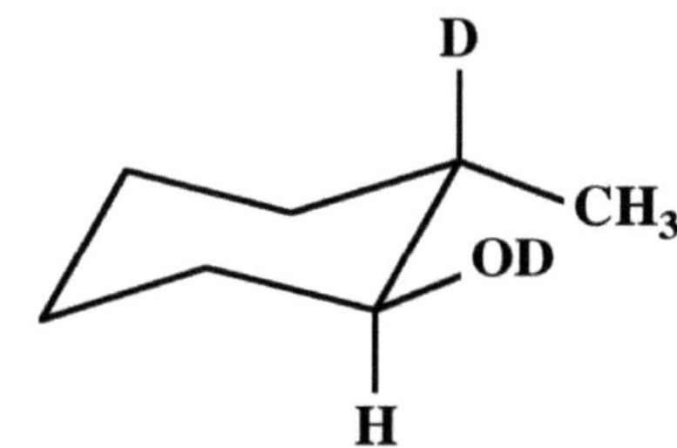

10. (24 points)

A $CH_3CHBrCH_3$

B

O

C

OH

D

11. (20 points)

E

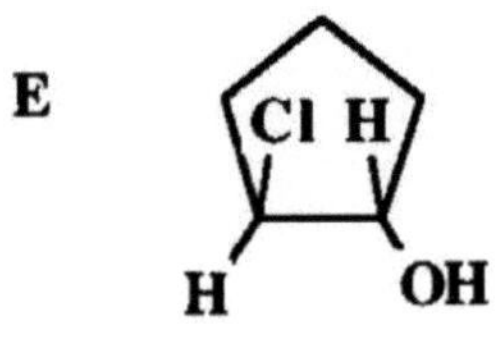

(racemic)

F

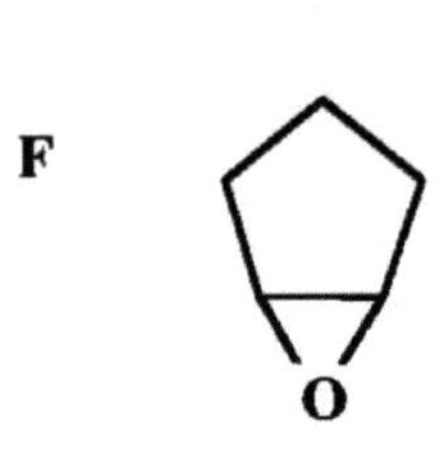

G

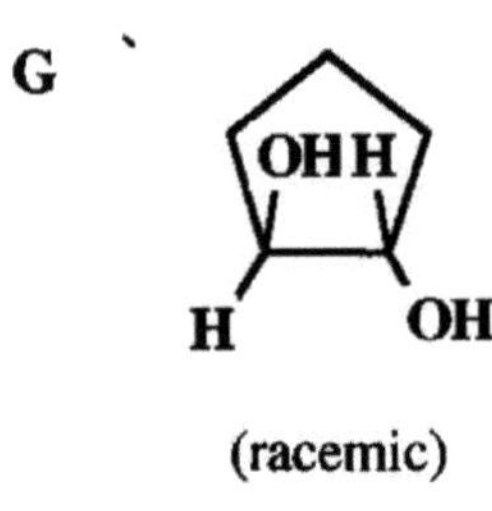

(racemic)

ORGANIC CHEMISTRY I FINAL EXAMINATION (II)
ANSWERS

1.

(a) Add each to a solution of potassium permanganate in aqueous base. The 1-hexene will react and cause the disappearance of the purple color while the cyclohexane will not react and the purple color will remain. Alternatively, each could be added to a solution of bromine in carbon tetrachloride. Again the 1-hexene would react to cause the loss of the purple color while the cyclohexane would not react and the color would remain.

(b) Add each to a solution of potassium permanganate in aqueous base. The pentanal would react causing the disappearance of the purple color while the 3-pentanone would not react, allowing the color to remain.

(c) Add each to a solution of potassium permanganate in aqueous base. The cyclopentanol would react causing the disappearance of the purple color while the 3-pentanone would not react, allowing the color to remain.

2.

(a)
3,3-dimethyl-2-butanol
(b)
trans-3-hexene
(c)
(*R*)-2-pentanol
(d)
1-bromo-2,2-dimethylpropane

3.

(a)
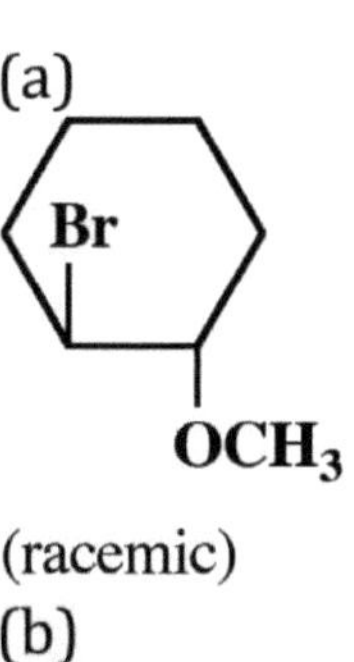

(racemic)
(b)
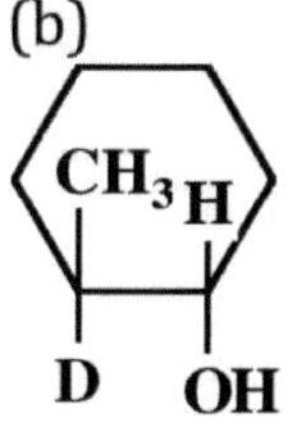

(racemic)
(c)
$K^{+}\ {}^{-}O_2C$ $CO_2^{-}\ K^{+}$

(d)
2-butanone
(e) 1,2-dimethylcyclohexene treated with H_2/PtO_2.

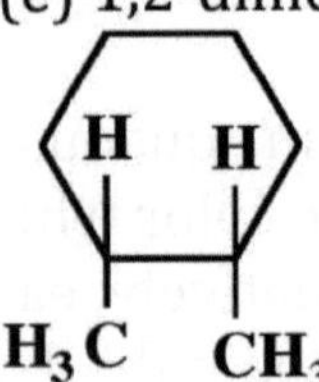

4.

Carbocations are stabilized by the delocalization of charge to the empty *p* orbital on the carbocation site. In such instances of simple carbocations, these electrons can come from the bonding molecular orbitals of adjacent C-H bonds by interaction as shown below:

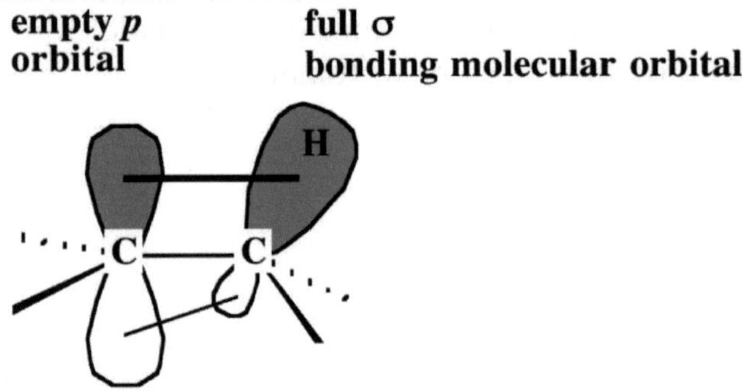

The more of these types of interactions present with a carbocation, the greater its stability. With a methyl cation, there are *none* of these interactions, while there are *nine* of these interactions with the *tert*-butyl cation, rendering the *tert*-butyl cation more stable.

5.

$$(CH_3)_2C{=}CH_2 + H\text{-}SO_3H \longrightarrow (CH_3)_3C^+ + HSO_4^-$$

$$(CH_3)_3C^+ + H_2O \longrightarrow (CH_3)_3C\overset{+}{O}H_2$$

$$(CH_3)_3C\overset{+}{O}H_2 + H_2O \longrightarrow (CH_3)_3COH + H_3O^+$$

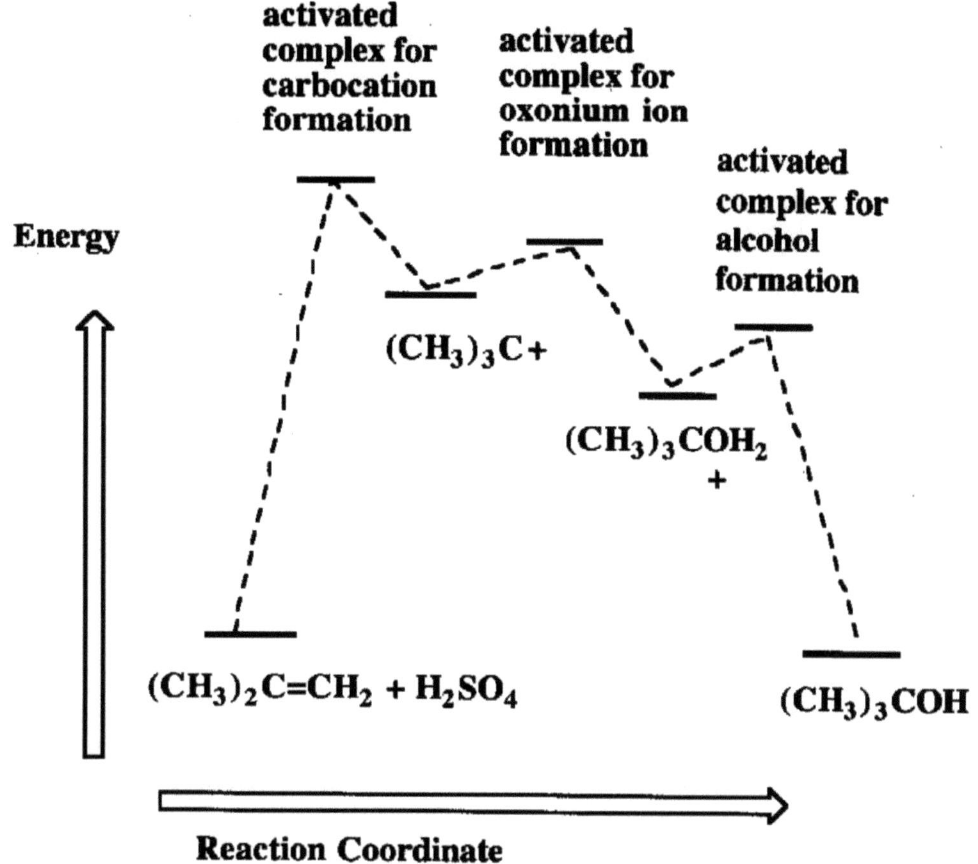

6.

A Br_2, light

B $K^{+-}OC(CH_3)_3$

C 1. BH_3 2. H_2O_2, KOH, H_2O

D CrO_3, H_2SO_4, H_2O

E 1. $Hg(CF_3CO_2)_2$, CH_3OH 2. $NaBH_4$

7.

The effect of the *tert*-butyl group is to lock the cyclohexane ring into a particular chair conformation. For the *trans* compound, shown at the bottom (below), the chlorine atom is located in an equatorial position leaving it with a dihedral angle of 60° with regard to all adjacent C-H bonds and thereby out of position for a facile E2 reaction. However, with the *cis* compound, shown at the top (below), the chlorine atom is in an axial position with a dihedral angle of 180° relative to two adjacent C-H bonds allowing it to undergo facile E2 reaction.

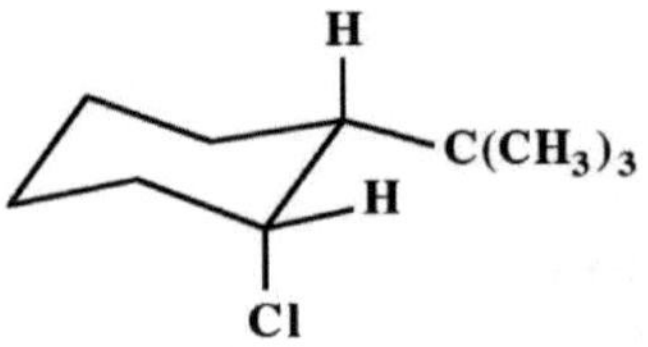

H

C(CH3)3 *cis*

H

Cl

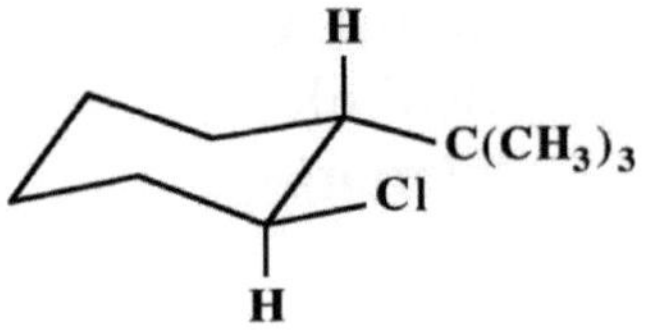

H

C(CH3)3 *trans*

Cl

H

8.

% composition -	C	62.04
	H	10.41
	O	27.55 (by difference from 100%)

C	62.04/12.011 = 5.16
H	10.41/1.008 = 10.32
O	27.55/15.999 = 1.72

Dividing each by 1.72, the empirical formula derived is C_3H_6O, which is also the molecular formula since the molecular weight is 58. With no hydroxyl group, but yet oxidizable, the structure deduced is that of propanal, CH_3CH_2CHO.

9.

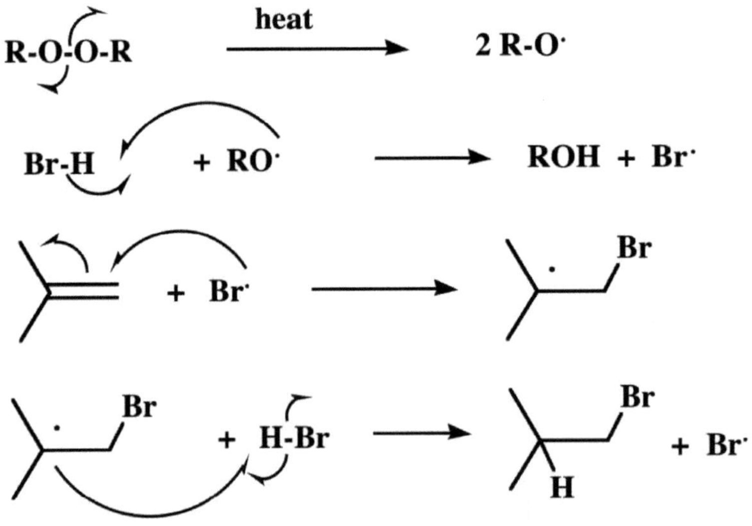

R-O-O-R —heat→ 2 R-O·

Br-H + RO· → ROH + Br·

+ Br· → Br

Br + H-Br → Br H + Br·

10.

(a) 1. H_2SO_4 2. HBr, peroxide, heat 3. $NaOCH_3$
(b) 1. $CrO_3/H_2O/H_2SO_4$ 2. CH_3CH_2MgI/diethyl ether 3. dilute aq. Acid
(c) 1. PBr_3 2. Mg/diethyl ether 3. D_2O

ANSWERS

1.

(a) 3,4-diethyl-3-hexene
(b) cycloheptene
(c) *trans*-2-butene
(d) 3-bromo-3-methylpentane

2.

(a) *trans*-1-bromo-2-ethoxycyclopentane (racemic)
(b) 1-deuterio-1-ethylcyclopentane
(c) 3-bromo-3-methylpentane
(d) *cis*-1,2-dimethylcyclopentane
(e) racemic 2-iodopentane

3.

(a) Add each to a dilute solution of potassium permanganate in aqueous base. Only the primary alcohol will decolorize the purple solution.

(b) Add each to a solution of bromine in carbon tetrachloride. Only the alkene will decolorize the solution.

(c) Add each to a dilute solution of potassium permanganate in aqueous base. Only the aldehyde will decolorize the solution.

4.

The filled 2*p* orbital on the oxygen of the methoxide ion interacts with the empty σ^* molecular orbital associated with the C-Br linkage. The major lobe of this latter orbital is on the side of the carbon opposite to that occupied by the Br, resulting in inversion of configuration about the carbon upon reaction of the methoxide ion. The product of the S_N2 reaction is the ether with configuration about the original electrophilic site being inverted.

5.

The tertiary bromoalkane, $(CH_3CH_2)_3CBr$, will react faster owing to the increased stabilization of the tertiary carbocation over the primary carbocation that would be formed in the alternative reaction to be considered.

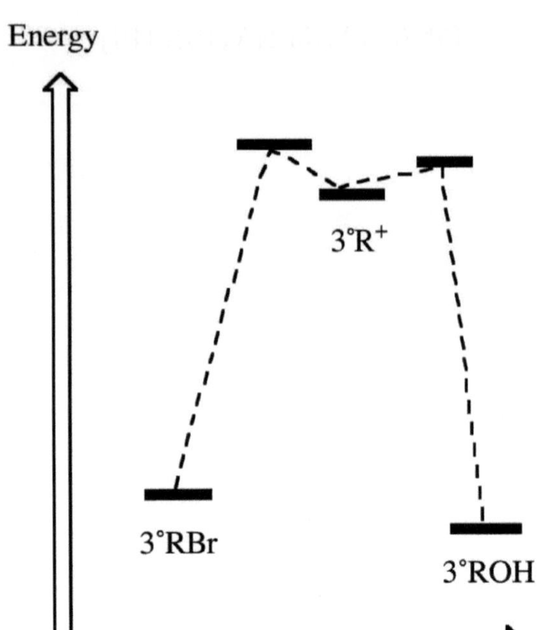

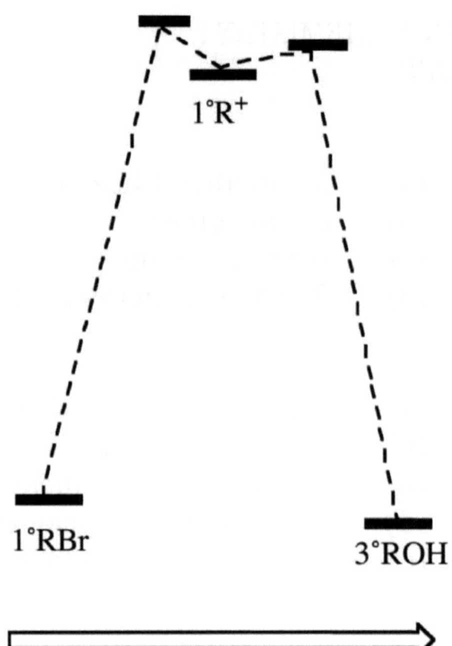

6.

A $C(CH_3)_3$

B H Br Br $C(CH_3)_3$

C $C(CH_3)_3$

D OCH_3 $C(CH_3)_3$

E OH H H $C(CH_3)_3$

7.

(a)

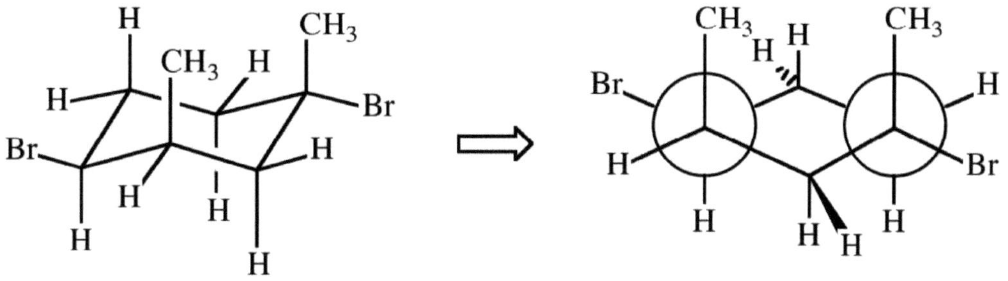

(b)

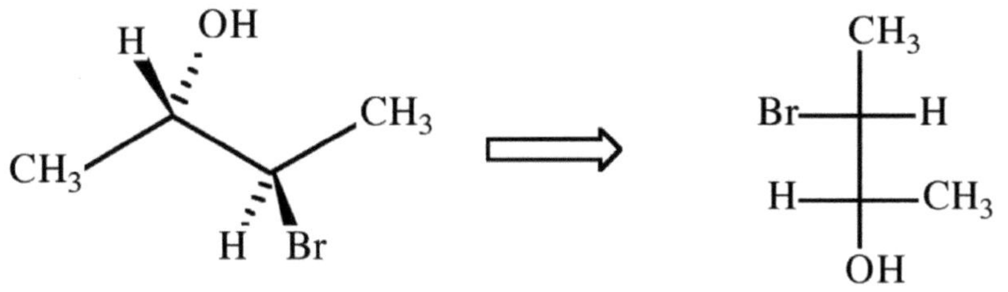

8.

With the *cis* compound the Br is in an axial position and thereby in good position for attack of a nucleophile to displace it. With the *trans* compound the Br is in an equatorial position and thereby must flip to an unfavorable conformation for substitution to occur. Thereby, the *cis*-compound will react more readily.

9.

(a) 1. PBr_3 2. Me, diethyl ether 3. D_2O
(b) 1. $CrO_3/H_2O/H_2SO_4$ 2. CH_3CH_2MgI 3. H_2O
(c) 1. Potassium *tert*-butoxide 2. $KMnO_4/KOH/H_2O$/heat 3. aq. acid

10.

(a) The separation of a pair of enantiomers into each of the pure enantiomeric forms
(b) A Lewis base acting to donate a pair of electrons to an atom other than hydrogen
(c) The lowest energy unoccupied molecular orbital in a given molecule
(d) Elimination of a pair of atoms or groups from between adjacent bound atoms

ORGANIC CHEMISTRY II
ANSWERS

FIRST EXAMINATION (I)

1. (20 points)

cyclohexene $\xrightarrow[\text{formic acid}]{H_2O_2}$ OH, OH (racemic) **A** $\xrightarrow[\text{heat}]{H_2SO_4}$ **B**

B $\xrightarrow{CH_3O_2C\text{-}C{\equiv}C\text{-}CO_2CH_3}$ **C** (CO_2CH_3, CO_2CH_3)

C $\xrightarrow{\text{heat}}$ **D** (CO_2CH_3, CO_2CH_3)

2. (20 points)

(a) Treat the starting alkene with benzyl alcohol in the presence of mercuric trifluoroacetate, followed by treatment with sodium borohydride in aqueous base

(b) Treat the starting alcohol with a base such as sodium hydride to generate the alkoxide ion, followed by benzyl bromide.

(c) Treat the 1-pentyne with sodium amide to generate the resulting acetylide ion, followed by iodomethane, and then do a sodium in liquid ammonia reduction of the alkyne to the *trans* alkene.

(d) Treat the benzyl ether with HBr in acetic acid.

3. (20 points)

The pentadienyl anion has its lowest energy three π orbitals filled, π_1, π_2 and π_3. The anion has the same π stabilization energy as does the radical and the cation because the last two electrons, those distinguishing the anion from the cation and the radical, are located in the π_3 molecular orbital, a nonbonding molecular orbital.

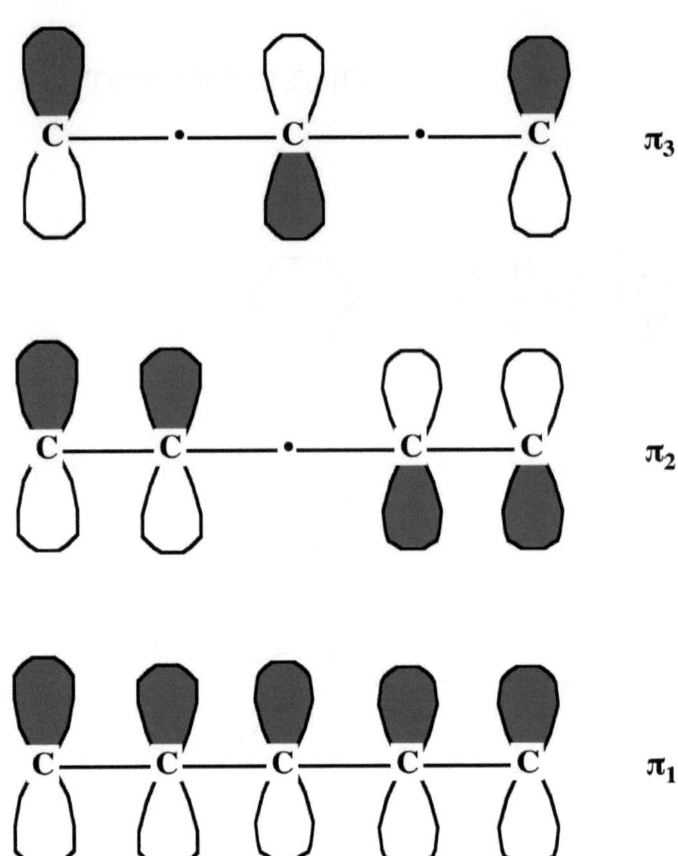

4. (20 points)
   (a) 0.9 δ, 3H, triplet
   1.0 δ, 3H, doublet
   1.3 δ, 2H, doublet of quartets
   1.9 δ, 1H, triplet of quartets
   2.5 δ, 1H, singlet
   (b)
   2.6 δ, 4H, doublet
   2.8 δ, 4H, singlet
   5.0 δ, 2H, triplet
   7.5 δ, 10H, broad
   (c)
   7.3 δ, 7H, singlet

5. (20 points)

F 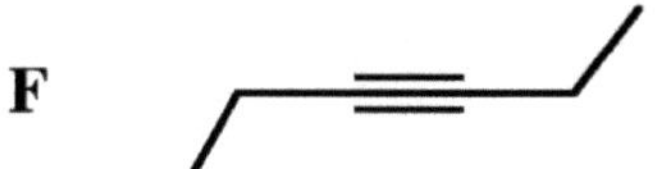

G 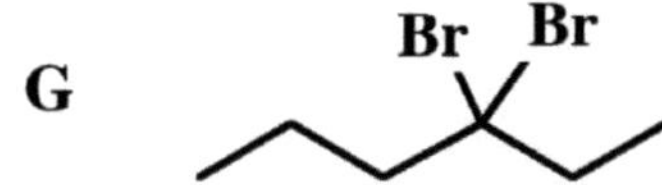

H 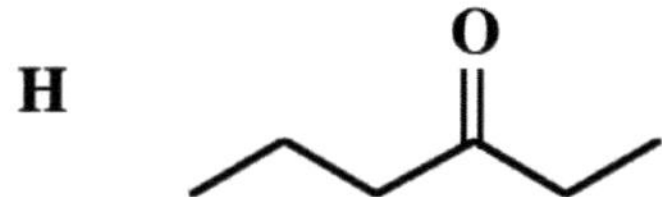

I

J

1. (20 points)

(a)

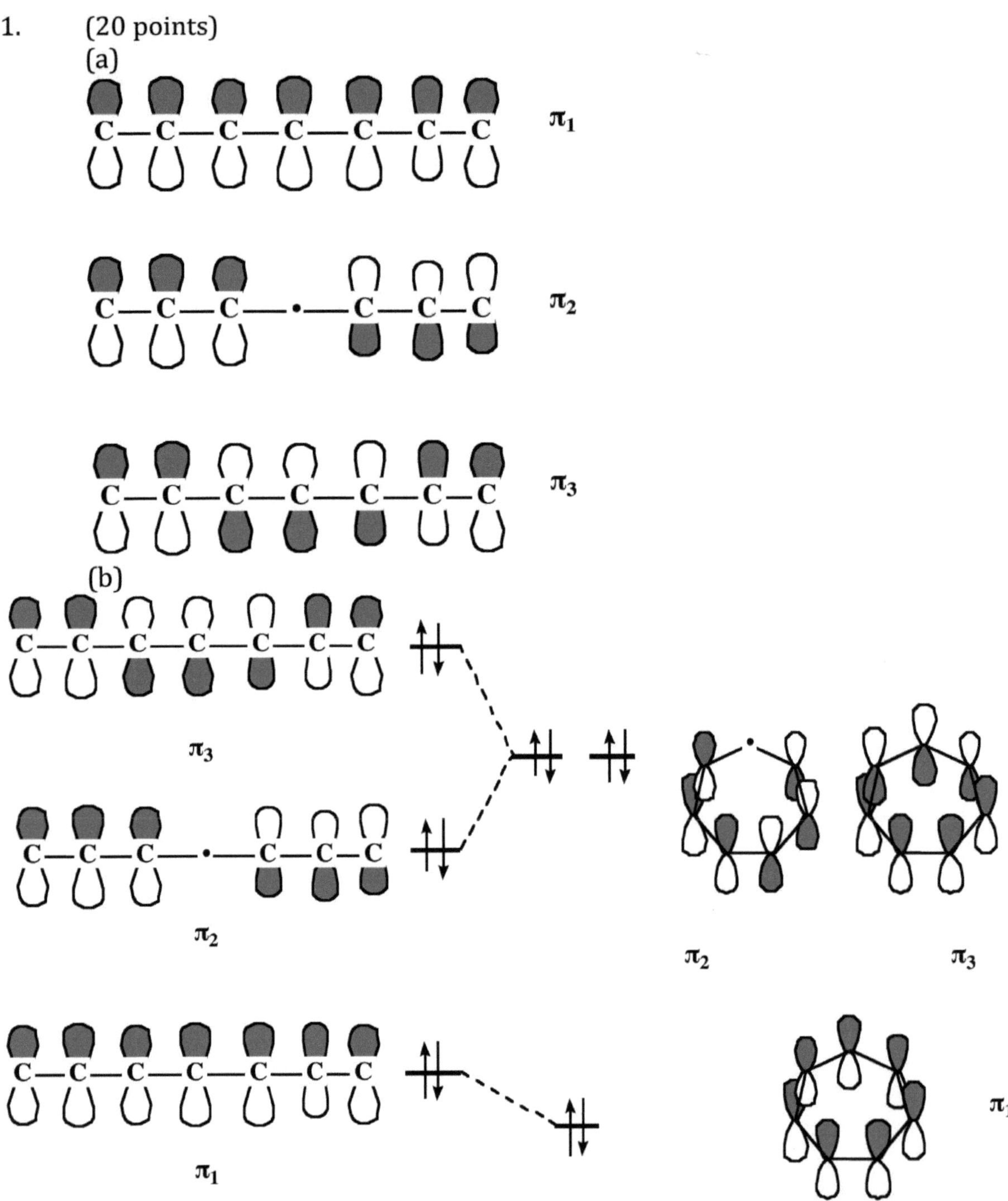

(c) There is no difference in π bonding stabilization for the heptatrienyl radical and the heptatrienyl anion (further, they have the same π bonding stabilization as does the heptatrienyl cation). The additional π electrons (beyond those of the heptatrienyl cation) are located in a nonbonding level for which neither stabilization nor destabilization is associated.

2. (20 points)

(a) *cis*-1,2-diiodocyclohexane

(b) 3-ethyl-2-pentene

(c) cyclohexyl methyl ketone

3. (20 points)

(a)

(i) For the diisopropyl ether the signal for the CH (attached to the O) would occur downfield (~ 3.5 δ) whereas for the 2,3-dimethylbutane it would occur relatively upfield (~ 1.3 δ).

(ii) Diisopropyl ether would exhibit a strong absorption for the C-O linkage at ~ 1100 cm^{-1}.

(b) The methyl propionate would exhibit a singlet relatively far downfield (~ 3.5 δ) whereas with the ethyl acetate the downfield signal would be a triplet with the singlet upfield (~ 2.0 δ).

4. (20 points)

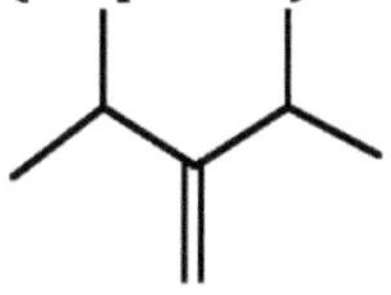

OCH_3 Br

A 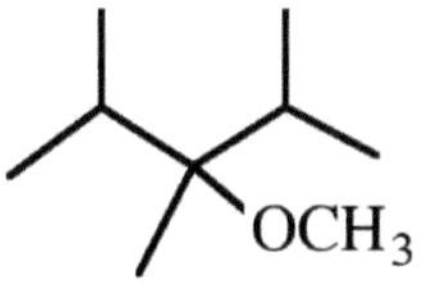 **B** 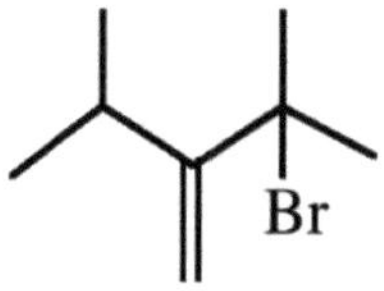 **C**

O O O

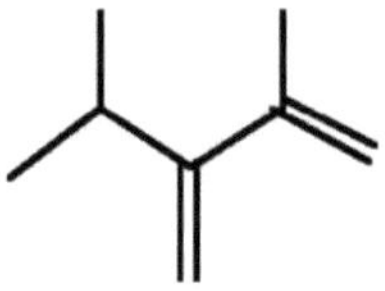

D 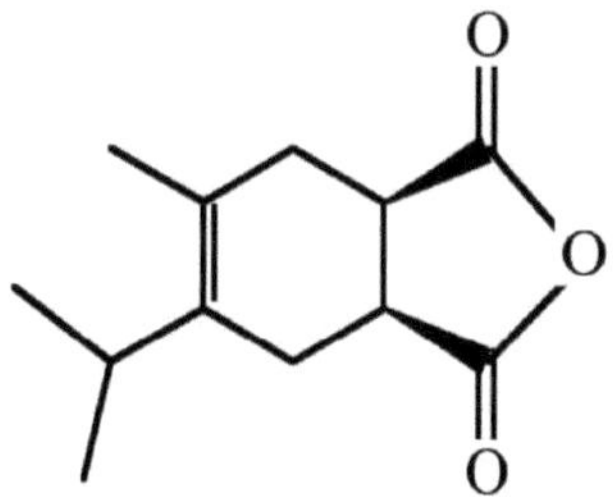 **E**

5. (20 points)

(a) Treat the cyclohexene with hydrogen peroxide in acetic acid to generate the corresponding epoxide. Then, treat the epoxide with methyl magnesium iodide and work up with water addition to give the target product.

(b) Treat the acetone with phosphorus trichloride to generate the 2,2-dichloropropane. Treat the 2,2-dichloropropane with sodium amide to double dehydrohalogenate it to propyne. Treat the propyne with sodium amide to generate the anion and add 1-bromopropane to the anion to give the target material.

(c) The (4*R*,5*S*)-dibromooctane is the *meso* structure for 4,5-dibromooctane, which requires *anti* addition of bromine to the *trans*-4-octene. The *trans*-4-octene is produced from the 4-octyne by reaction with sodium in liquid ammonia.

1. (15 points)

A 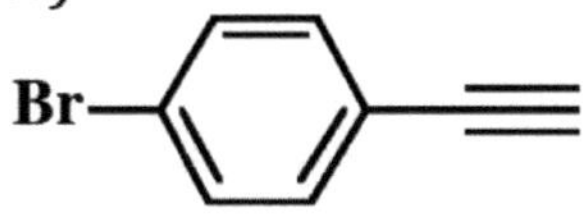 Br

B 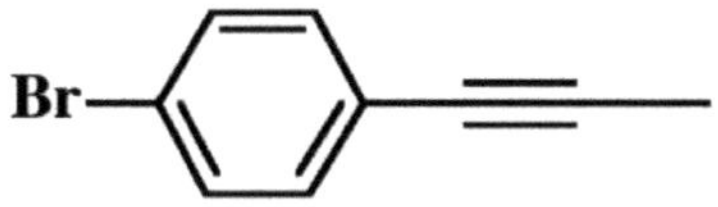 Br

C Br

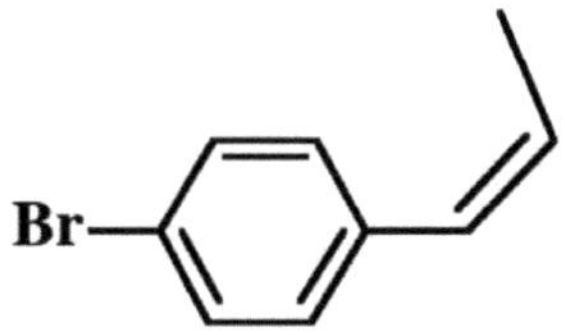

2. (10 points)

(a)

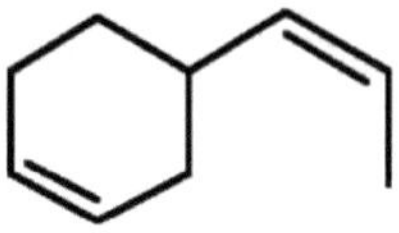

(b)

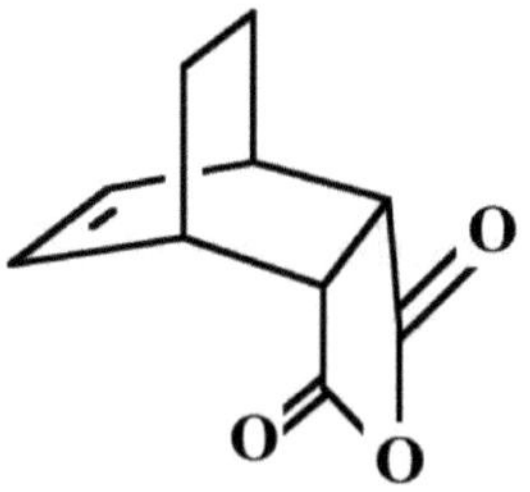

O

O O

3. (14 points)

(a) Treat the starting alkene with bromine in carbon tetrachloride followed by double dehydrohalogenation using an excess of sodium amide. The resulting anion of the alkyne is then treated with iodoethane to give the internal alkyne which is reduced with sodium in liquid ammonia to give the target alkene.

(b) The starting alkyne is treated with sodium amide to generate the anion which is then treated with iodoethane to generate the symmetrical alkyne. The symmetrical alkyne is then hydrated with mercuric trifluoroacetate in aqueous acid to give the 3-hexanone which is reduced with sodium borohydride to the target alcohol.

4. (20 points)

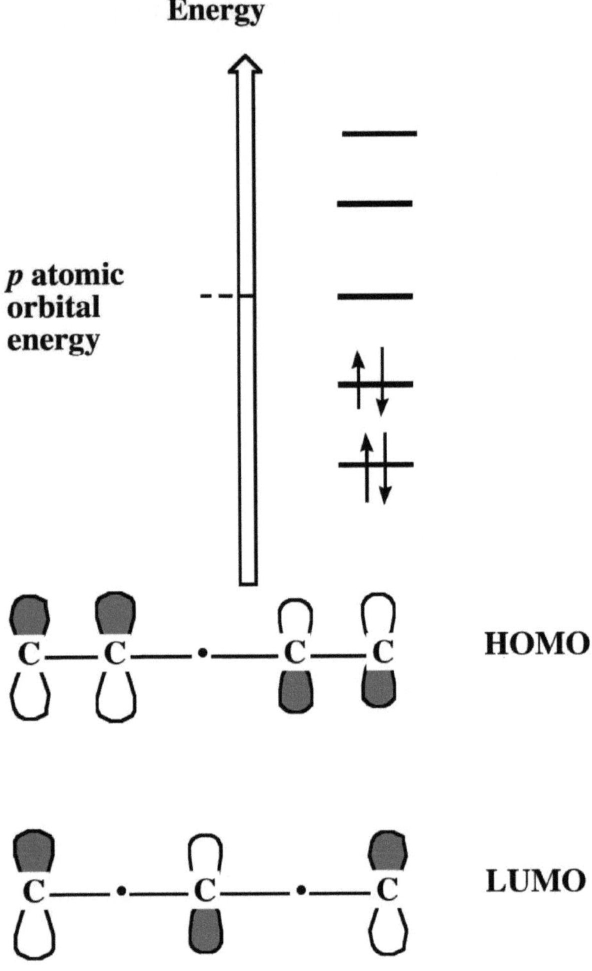

Adding two electrons would not change the π bonding energy of this system. The additional two electrons would be placed in the nonbonding level which would not change the π bonding stabilization.

5. (20 points)

(a) It is not a cyclic π system - there is not a *p* orbital at each atomic site around the entire ring.

(b) It is not a cyclic system.

(c) It is a free radical. There are ($4n$+1) electrons in the system, the last electron being located in an antibonding orbital.

(d) The central hydrogen atoms bump into each other distorting the system from planarity and keeping the π bonds from interacting properly to be aromatic.

6. (21 points)

(a) 1.3 δ, 4H, doublet of triplets
1.6 δ, 1H, quintet
2.5 δ, 1H, singlet
3.5 δ, 4H, triplet of triplets

variable chemical shift depending on concentration and temperature, 2H, triplet

(b) 2.2 δ, 4H, doublet
5.0 δ, 2H, triplet
12.0 δ, 2H, singlet

(c) 2.3 δ, 3H, singlet
the remaining hydrogens would all be exchanged with D from the solvent and appear at 4.2δ, the chemical shift for water.

1. (25 points)

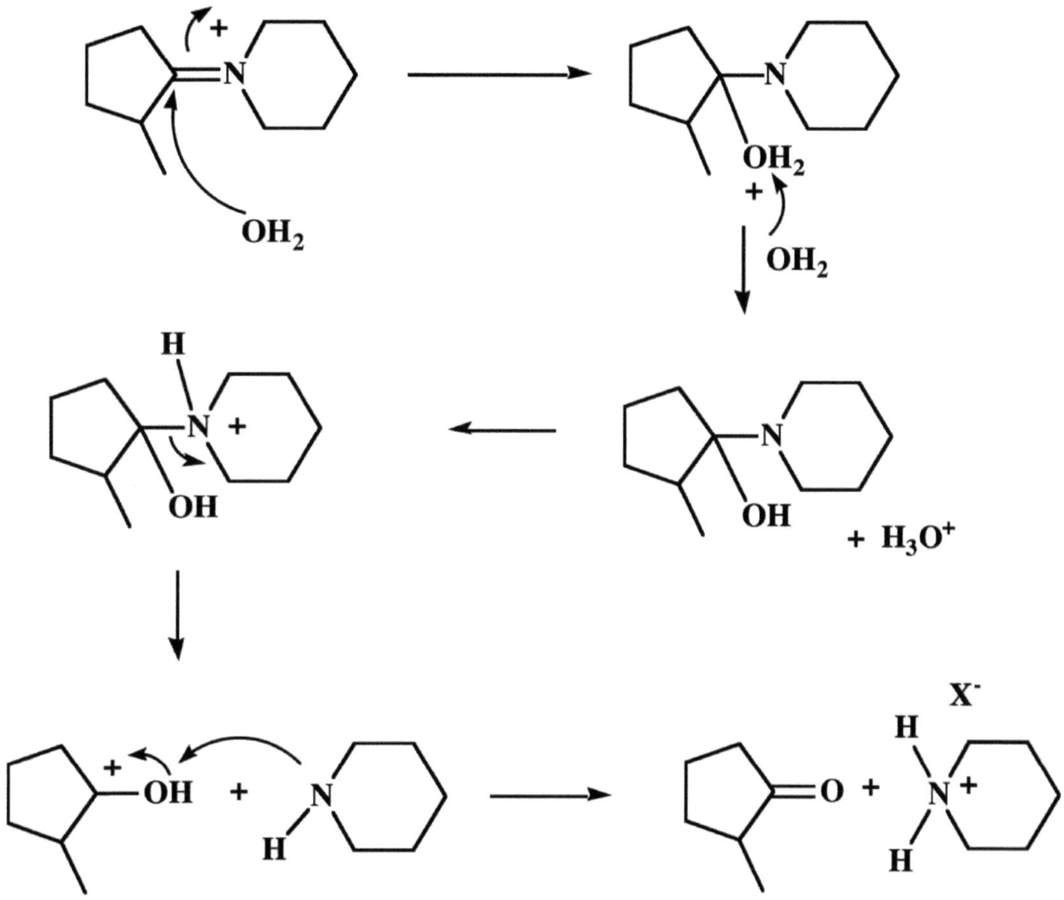

+ H_3O^+

X^-

2. (27 points)

(a)

CH_3CH_2O–C(=O)–CH_2–C(=O)–OCH_2CH_3

1. NaOEt
2. CH_3CH_2I
3. aq. acid, heat

(b)

H_3C–C(=O)–CH_2–C(=O)–OCH_2CH_3

1. NaOEt
2. $BrCH_2CH_2CH_2CH_2Br$
3. NaOEt
4. aq. acid, heat

(c)

H_3C–C(=O)–CH_2–C(=O)–OCH_2CH_3

1. Product from (b), $SOCl_2$
2. Product from (a), $NaHCO_3$
3. combine from steps 1 and 2

3. (20 points)

This is an intramolecular transesterification. An inert solvent should be used (*e.g.*, toluene) with an anhydrous acid catalyst (*e.g.*, *p*-toluenesulfonic acid), and the reagent (the methyl 5-hydroxypentanoate) should be dilute in this medium to minimize *intermolecular transesterification*. The reaction system needs to be heated sufficiently to distill the methanol as it is produced and thereby be removed from the reaction medium. If the reaction mixture is not heated sufficiently, allowing the methanol to remain in the reaction system, reaction will be incomplete, and if the methyl-5-hydroxypentanoate is too concentrated, intermolecular reaction will occur.

4. (28 points)

A

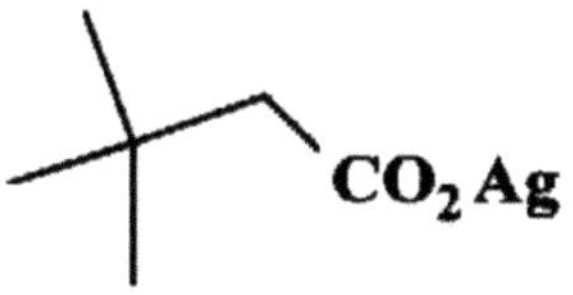

CO_2Ag

B

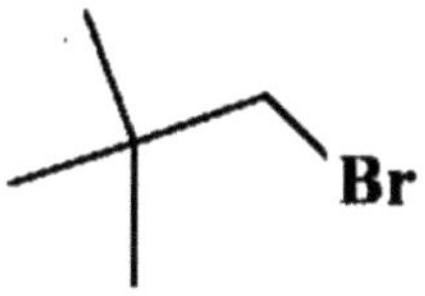

Br

C

NH_2

D

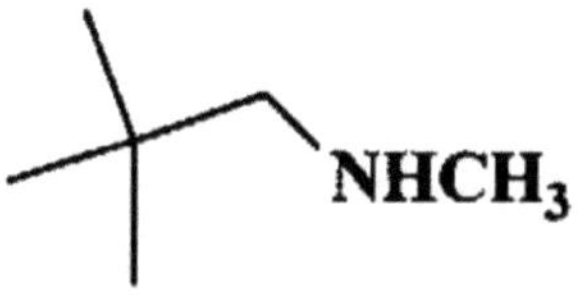

$NHCH_3$

E

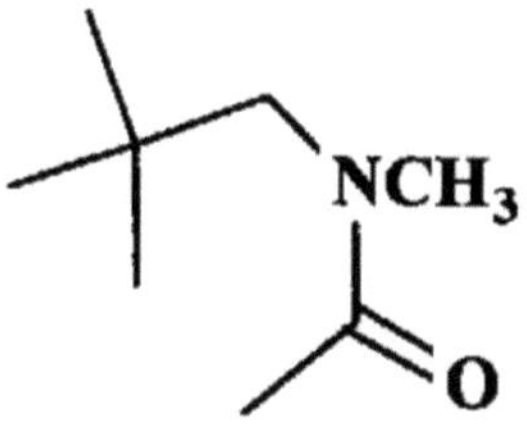

NCH_3

O

F

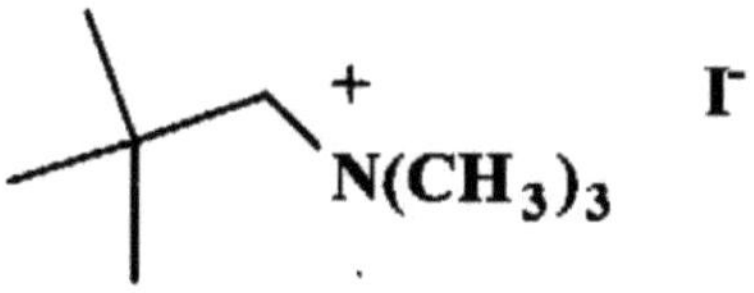

$\overset{+}{N}(CH_3)_3$ I^-

G

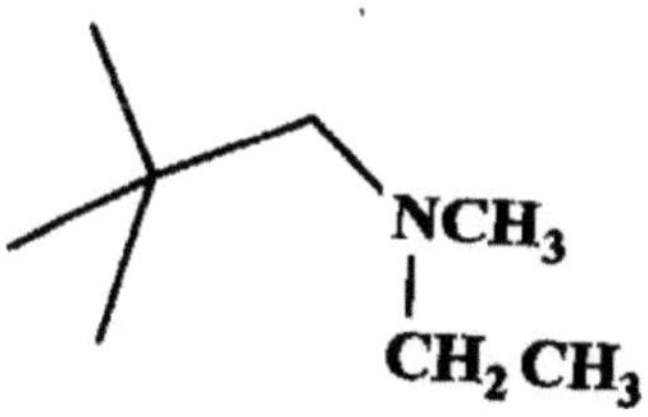

NCH_3

CH_2CH_3

1. (25 points)

(a) 1. Mg, ether
2. CO_2, aq. acid workup

(b) 1. Ag_2O
2. Br_2, heat

(c) 1. Cl_2, NaOH, H_2O
2. Br_2, PBr_3, aq. acid workup

(d) 1. H_2SO_4, HNO_3
2. Br_2, Fe
3. Sn/HCl
4. $NaNO_2$, H_2SO_4, CuCl

2. (27 points)

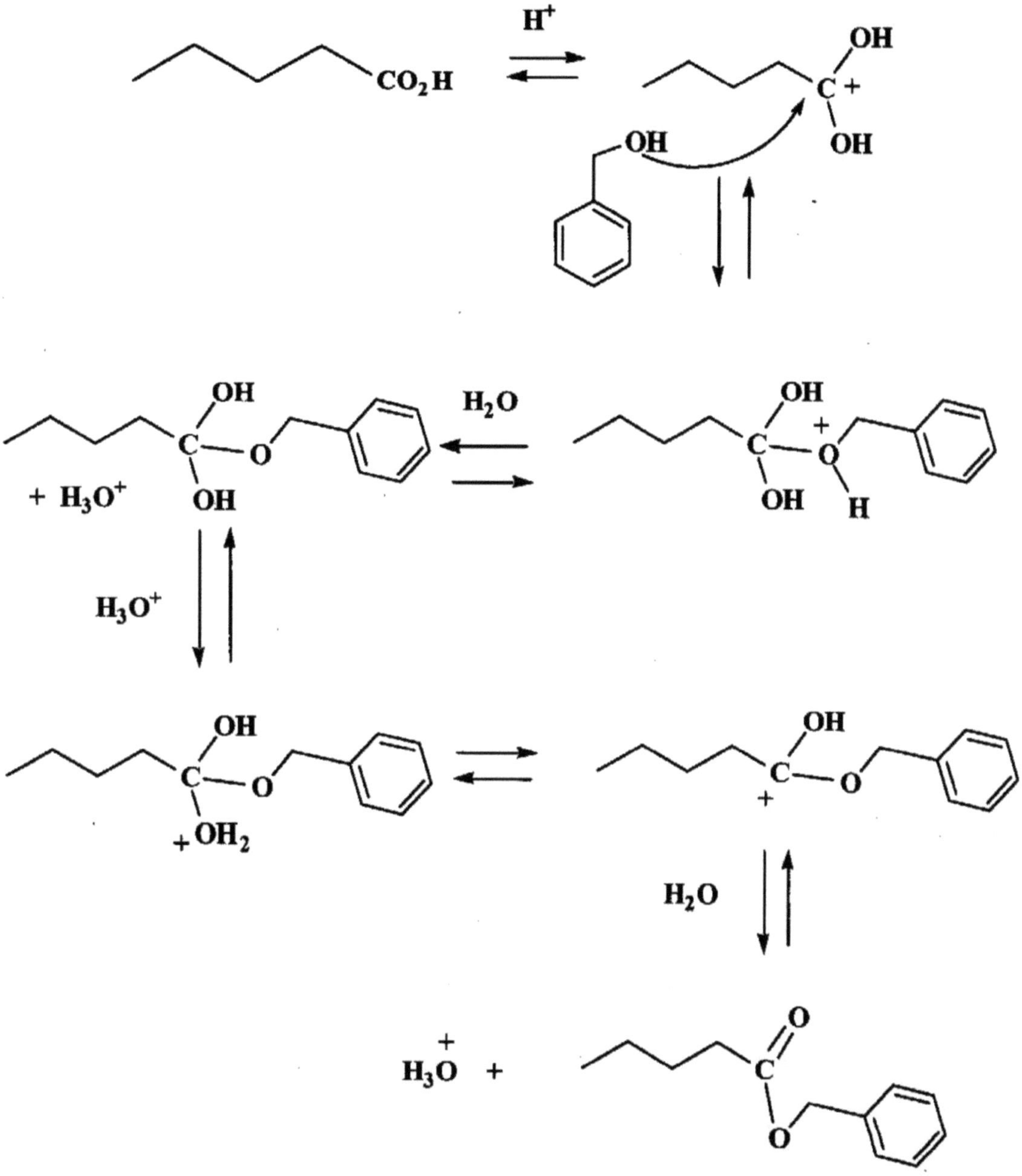

H^+

CO2H ⇌ (protonated acid, C+ with two OH)

(benzyl alcohol OH attacks C+)

OH, C—O(+)(H)—CH2Ph, OH ⇌ H_2O ⇌ OH, C—O—CH2Ph, OH + H_3O^+

H_3O^+

OH, C—O—CH2Ph, $^+OH_2$ ⇌ OH, C+—O—CH2Ph

H_2O

$H_3\overset{+}{O}$ + C(=O)—O—CH2Ph

3. (20 points)

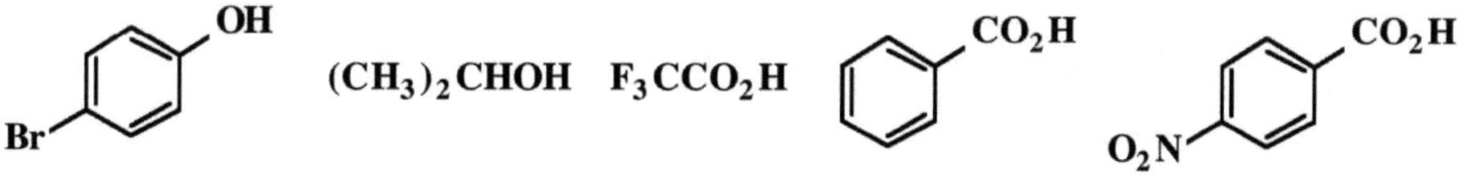

Br-C6H4-OH	$(CH_3)_2CHOH$	F_3CCO_2H	C6H5-CO2H	O_2N-C6H4-CO2H
2	1	5	3	4

4. (15 points)

(a) heptanoic acid, Br_2, PBr_3, aq. acid workup

(b) 1. heptanoic acid, Ag_2O
2. Br_2, heat

(c) 1. $SOCl_2$
2. $Li(CH_2CH_2CH_3)_2Cu$
3. Sn(Hg)/HCl

5. (30 points)

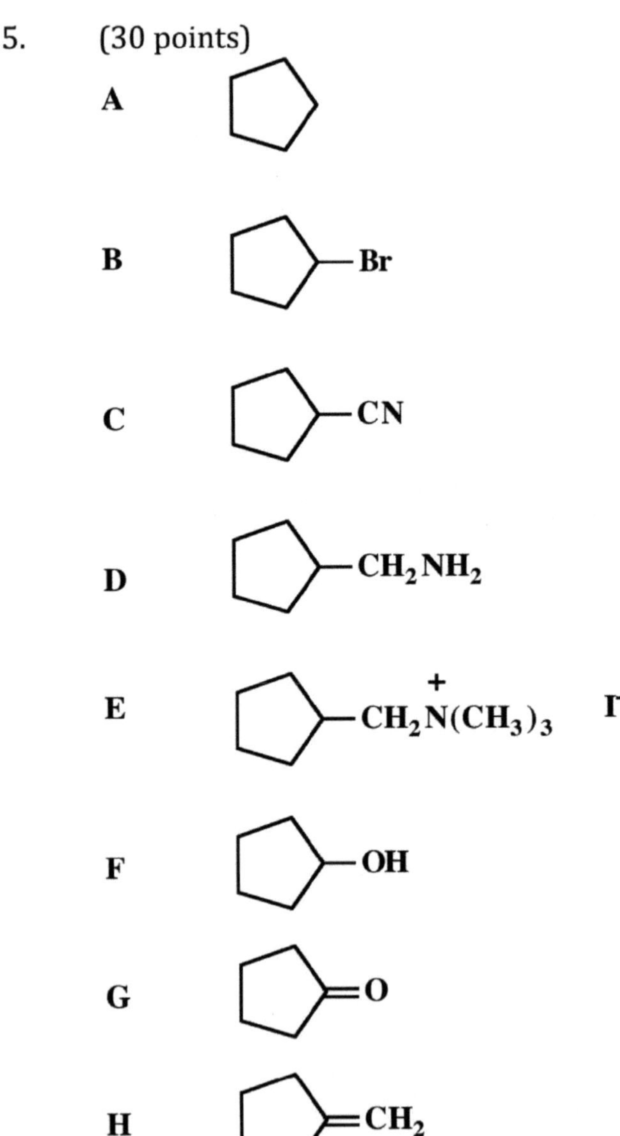

1. (20 points)

(a) 1. Mg, ether
2. CO_2
3. aq. acid workup

(b) 1. H_2SO_4, HNO_3
2. Cl_2, Fe
3. $NaNO_2$, H_2SO_4, CuCl

(c) 1. C_6H_5CHO, acid catalyst
2. H_2, PtO_2

(d) 1. $NaBH_4$, 2-propanol
2. PBr_3

2. (20 points)

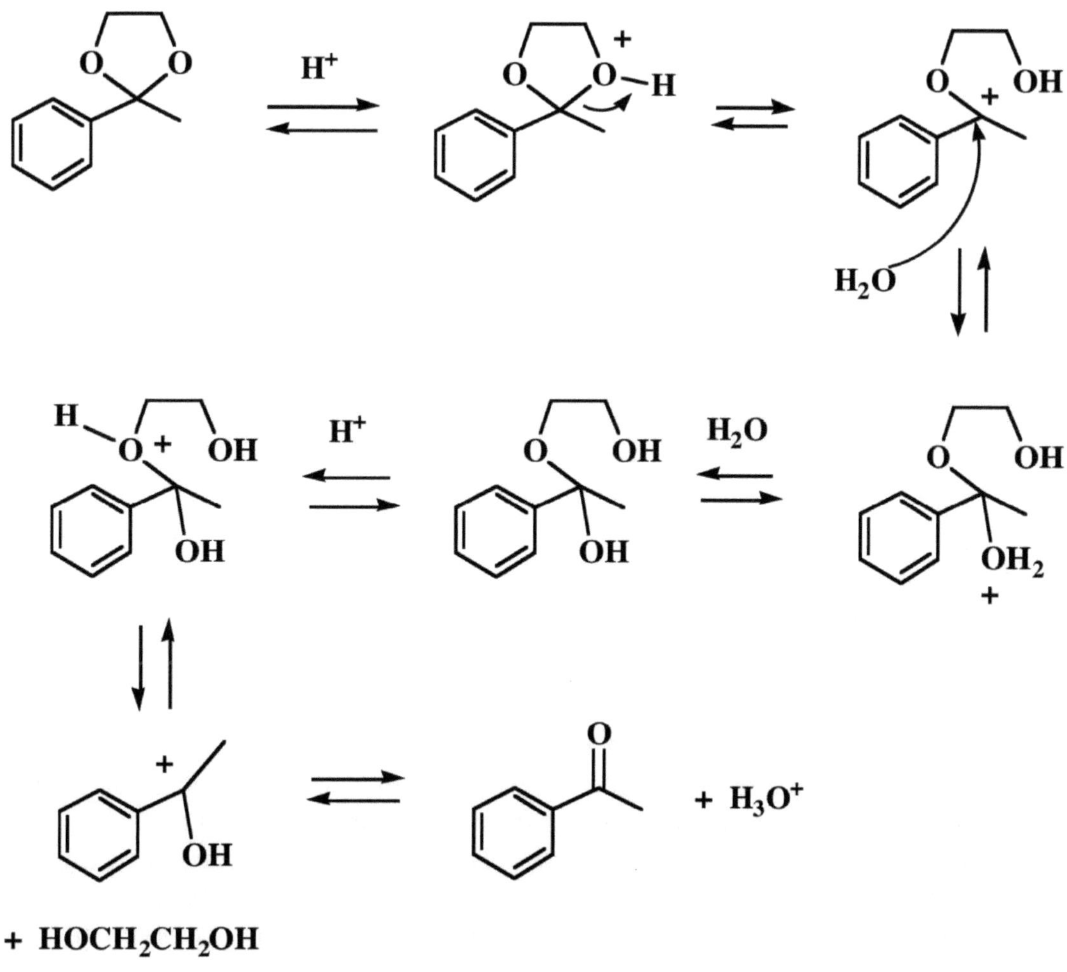

H^+ H_2O H^+ H_2O

+ H_3O^+

+ $HOCH_2CH_2OH$

3. (20 points)

A 

B CH_2OH

C CHO

D $CH{=}CH_2$

E O

4. (20 points)

Decreasing Basicity

N N H N(CH_3)$_2$ NO_2 N(CH_3)$_2$ N H

The tertiary amine with all alkyl groups out of the way of interacting with the acidic site is most basic, followed by the secondary amine. Basicity is related to the electron donating character of the alkyl groups. The simple tertiary aromatic amine follows, with the electron-withdrawing aromatic ring decreasing the basicity compared to the aliphatic amines. Putting a nitro group on the ring (an electron-withdrawing group) decreases the basicity, while with the pyrrole, protonation would destroy the aromatic character and thereby pyrrole is least basic of the group.

5. (20 points)

(a)

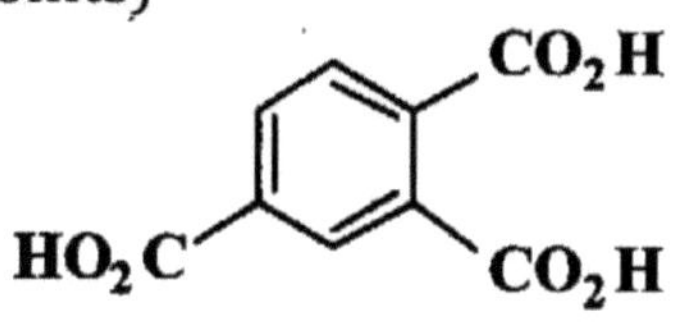

HO_2C CO_2H CO_2H

(b)

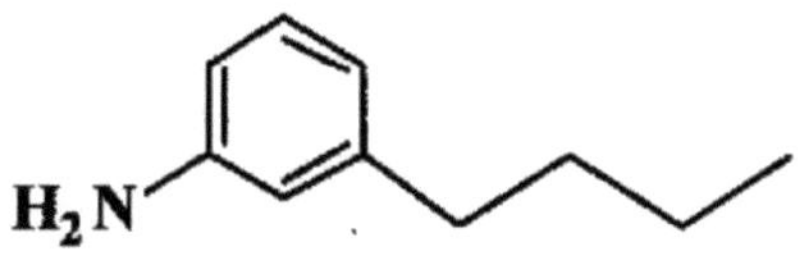

H_2N

(c)

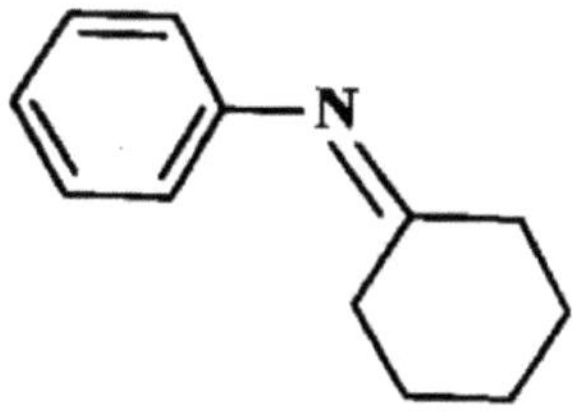

N

(d)

O=N— —$N(CH_3)_2$

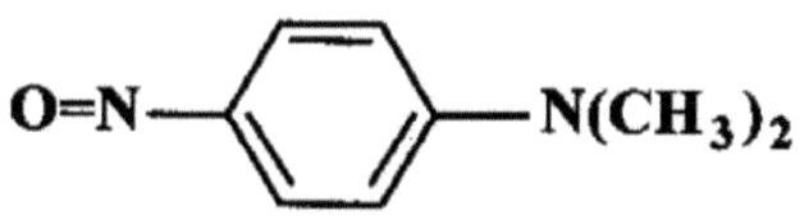

1. (20 points)

I $NaIO_4$

II $(C_6H_5)_3PCH_2$

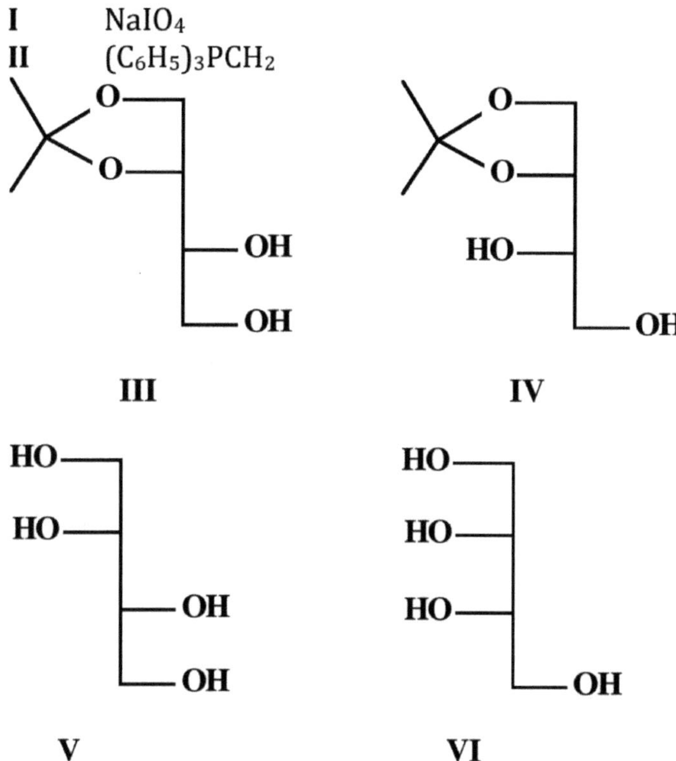

2. (20 points)

(a) We need to protect the amino nitrogen of alanine so that when we activate its carboxyl group for coupling to the protected glycine the activated (protected) alanine does not react with additional molecules of alanine to generate polyalanine.

(b) We need to protect the carboxyl group of glycine prior to allowing it to react with the activated (protected) alanine to prevent from occurring any reaction at the carboxyl group rather than specifically at its amino group.

(c) Isobutyl chloroformate will react with the amino nitrogen of our protected alanine to generate a type of anhydride linkage at the alanine carboxyl group, thereby imparting to it reactivity for coupling with an amino group of another amino acid molecule.

(d) Acetyl chloride *would* protect the amino nitrogen of alanine by forming an amide, but it would be such an amide that the conditions necessary to remove the protecting group would also be sufficient to cleave the desired amide linkage between the alanine and glycine units that we have been working to generate.

3. (20 points)

Direct treatment of cyclohexanone with ethyl iodide and sodium hydroxide would entail several difficulties for economical isolation of the target 2-ethylcyclohexanone. These include: (a) the possibility of multiple alkylation of cyclohexanone which would use additional ethyl iodide, would destroy in a useless reaction some of the cyclohexanone, and would cause significant difficulties for isolation of the target material from by-products; (b) direct reaction of the ethyl iodide with sodium hydroxide producing ethanol and destroying reagent ethyl iodide in a useless process; (c) the possibility of producing a direct aldol-type condensation of the cyclohexanone, destroying reagent cyclohexanone in a useless reaction and making the final isolation of the target material from undesirable by-product more difficult. Although the use of an enamine synthesis involves two steps and uses piperidine, the reaction is controlled by this approach such that undesirable by-products are not formed and isolation of the target product is simplified. Further, the piperidine is recyclable and may be recovered and used in further preparations.

4. (20 points)

$$CH_2(CO_2CH_2CH_3)_2 \xrightarrow[CCl_4]{Br_2} BrCH(CO_2CH_2CH_3)_2$$

$$BrCH(CO_2CH_2CH_3)_2 + \text{(potassium phthalimide, } :\overset{-}{N}:\ K^+\text{)} \longrightarrow \text{Phth}N{-}CH(CO_2CH_2CH_3)_2$$

$$\text{Phth}N{-}CH(CO_2CH_2CH_3)_2 \xrightarrow[2.\ C_6H_5CH_2Br]{1.\ NaOCH_2CH_3} \text{Phth}N{-}C(CO_2CH_2CH_3)_2(CH_2C_6H_5)$$

$$\text{Phth}N{-}C(CO_2CH_2CH_3)_2(CH_2C_6H_5) \xrightarrow[2.\ \text{aq. acid, heat}]{1.\ NaOH/H_2O} H_2N{-}CH(CH_2C_6H_5)CO_2H$$

5. (20 points)

(a) H_2C-CH_2CHO with $CH(CO_2CH_3)_2$ on the first carbon

(b) 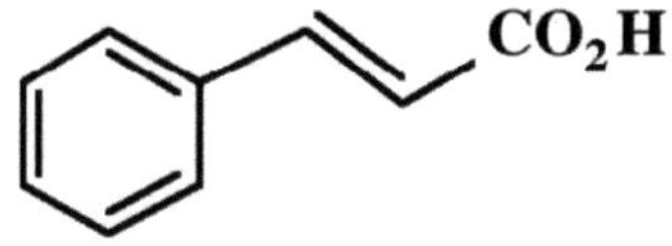 CO_2H

(c) 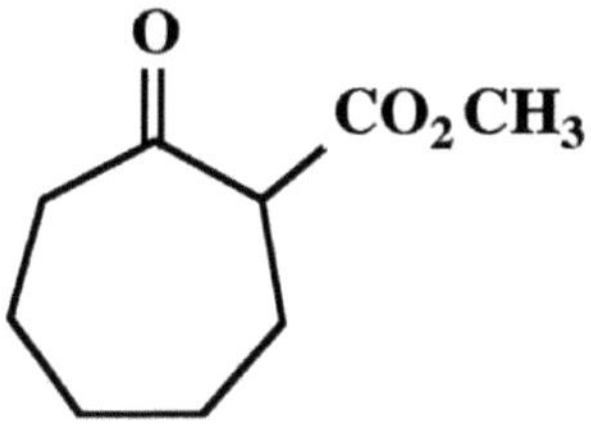 O CO_2CH_3

(d) $H_2C{=}CHCO_2H$

1. (20 points)

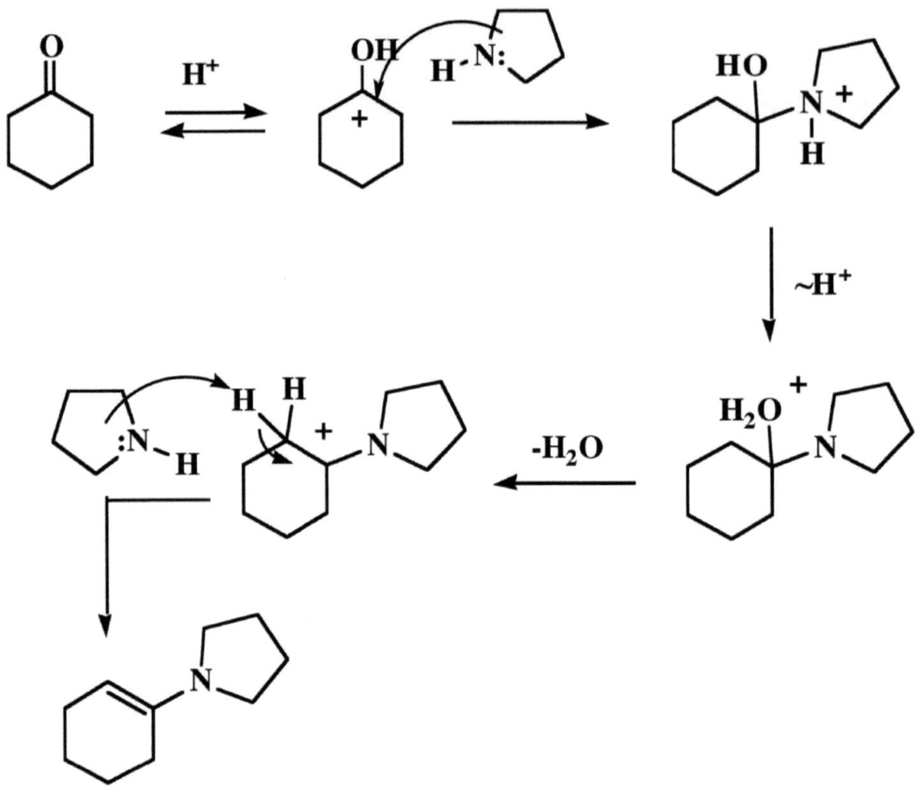

2. (15 points)

The difficulty arises because the student had the methyl propanoate in a relatively high concentration at the beginning of the experiment and there was a major competition for isolation of the desired product by the occurrence of an ordinary Claisen condensation of the methyl propanoate. To avoid this problem the student needed to mix the base and the benzaldehyde in the solvent, and slowly with vigorous stirring add dropwise the methyl propanoate to the reaction mixture. This will minimize the extraneous Claisen condensation of the methyl propanoate and allow isolation of a good yield of the target compound.

3. (24 points)

From the molecular weight and the elemental analysis, we conclude that the carbohydrate is a tetrose. Measurement of the ^{1}H NMR would differentiate the possibilities of it being an aldose or a ketose, through the presence (aldose) or absence (ketose) of a signal at ~10 δ. Should it be a ketose, the only differentiation needed would be the stereochemistry about the single stereogenic site. This differentiation could be accomplished by treatment of the material with Cl_2/aq. NaOH and comparison of the resulting carboxylic acid with that derived from D-glyceraldehyde upon treatment with aqueous ammoniacal silver nitrate by optical rotation. Were the unknown compound to have the D-configuration, the optical rotation of the product carboxylic acid would be the same as that of the acid from D-glyceraldehyde, and of the L-configuration if its optical rotation were opposite to that of the acid from D-glyceraldehyde.

If the carbohydrate were an aldotetrose, we would need to perform a Ruff degradation (or Wohl degradation) to generate glyceraldehydes and thereby determine (by comparison of optical activity with D-glyceraldehyde) if it were of the D- or the L-configuration. There would remain to determine if it were of the threose or the erythrose family. This could be accomplished by performing a reduction ($NaBH_4$) of the carbohydrate and checking the optical activity of the product. If it were of the erythrose family, it would have zero optical activity, whereas if it were of the threose family it would have a non-zero optical activity.

4. (21 points)

(a) Heat the furfural with acetic anhydride and sodium acetate to perform a Perkin condensation and isolate the target product.
(b) Heat the starting amide with Br_2 in aq. NaOH
(c) Reduce the starting amide with $LiAlH_4$, working up with water.

5. (20 points)

The enolate ion (a nucleophile) will add to the LUMO of the acrolein. As it is a conjugated system, the LUMO of the acrolein will be π_3^* for which the largest lobe is located on the β-carbon site. Thus, an incoming nucleophile will add preferentially at the β-carbon site rather than at the carbonyl carbon atom (which would also require disruption of the π bonding interaction within that molecular orbital).

1. (24 points)

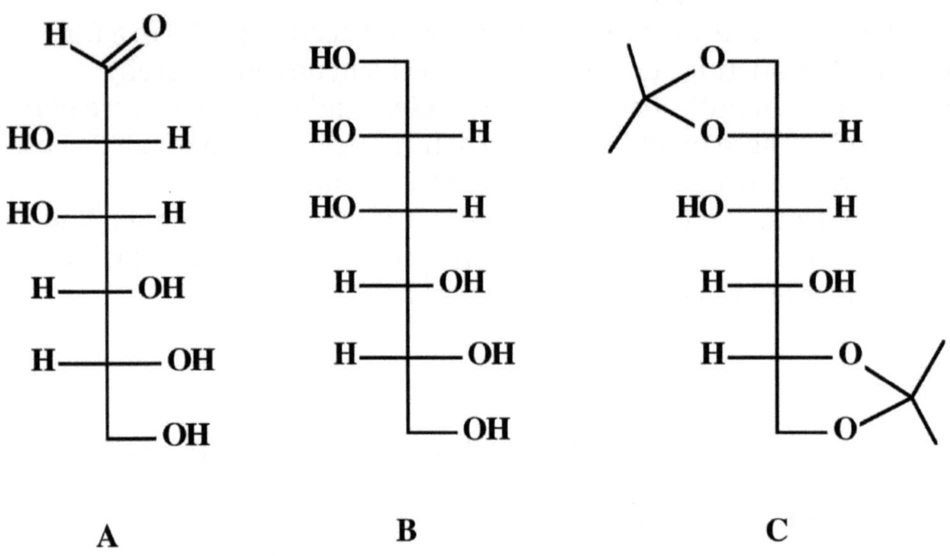

CH_2OH

$CH_2OC(O)CH_3$

D E F

2. (18 points)

The alanine produced in this synthesis is racemic. The reaction of the racemic alanine with the D-tartaric acid will produce a pair of diastereoisomeric salts, as shown below. As diastereoisomers, they will exhibit significantly different physical properties, including melting points and solubilities in a variety of solvent systems, such as methanol.

CO_2H / HO–H / H–OH / CO_2^- H_3C, H, CO_2H, NH_3^+ and CO_2H / HO–H / H–OH / CO_2^- H_3C, H, CO_2H, NH_3^+

3. (27 points)

(a) 1. Treat glycine with 1 equivalent of benzoyl chloride to generate hippuric acid

2. Treat hippuric acid with acetic anhydride to cyclize system to an azlactone
3. Treat the azlactone with NaOH and add benzyl chloride
4. Hydrolyze the system to generate the phenylalanine

(b) 1. Treat with Br_2 in acetic acid to monobrominate at the α-position.

2. Heat with ammonia to displace the bromide and introduce the amino group.

(c) 1. Heat with ammonia and NaCN

2. Heat with aqueous acid to hydrolyze the nitrile linkage and form the product.

4. (19 points)

1. Treat L-alanine with carbobenzyloxy chloride to protect the amino group
2. Treat L-phenylalanine with methylpropene and acid to protect the carboxyl group as a *t*-butyl ester.
3. Treat the protected L-alanine with isobutyl chloroformate to activate the carboxyl group for coupling.
4. Add to the activated/protected L-alanine the protected L-phenylalanine.
5. Heat with TFA to deprotect the material and generate the free Ala-Phe.

5. (12 points)

(a) The molecular ion is the species in a mass spectrometric experiment that is produced by the removal of a single electron from the parent molecule. It is represented by $M^{+\cdot}$ with a m/q ratio corresponding to the molecular weight of the compound.

(b) The molar absorptivity is the efficiency with which a molecule absorbs light of a particular wavelength. It is provided by the absorbance of a sample divided by the molar concentration of the sample in the experiment and by the length of the cell in cm.

(c) This is the region of the ultraviolet spectrum, of wavelength < 200 nm, within which molecules in the air (O_2, N_2, CO_2) absorb the UV light.

1.

(a) 1. 1-butanol treated with CrO_3/H_2SO_4 to generate butanoic acid
2. butanoic acid treated with 2-propanol and acid catalyst to give the target material

(b) 1. cyclopentanol treated with CrO_3/H_2SO_4 to generate cyclopentanone
2. cyclopentanone treated with ethanol and acid catalyst to generate the target material

(c) 1. 1-pentanol treated with PBr_3 to give 1-bromopentane
2. ethanol treated with PBr_3 to give bromoethane
3. bromoethane treated with sodium amide to give ethylene
4. ethylene treated with Br_2 to give 1,2-dibromoethane
5. 1,2-dibromoethane treated with sodium amide to give ethyne
6. ethyne treated with sodium hydride to give sodium acetylide
7. sodium acetylide treated with 1-bromopentane to give 1-heptyne
8. 1-heptyne treated with $HgSO_4/H_2SO_4/H_2O$ to give the target material

2.

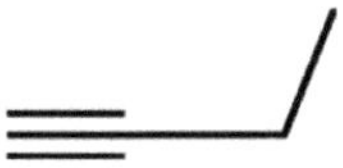

A

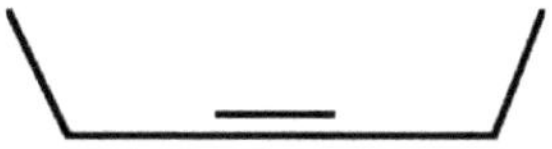

B

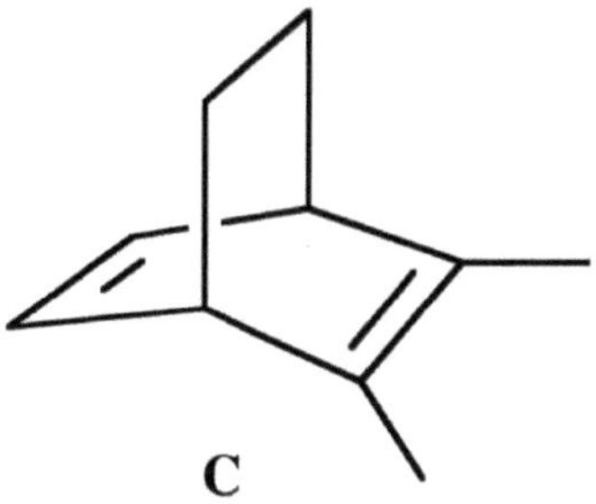

C

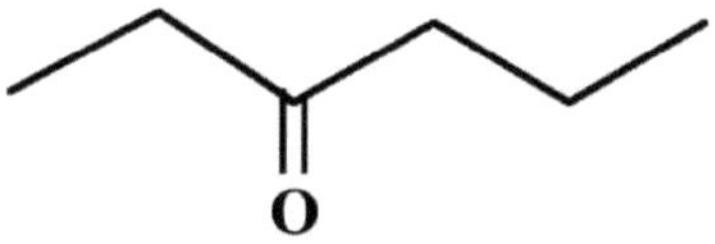

O

D

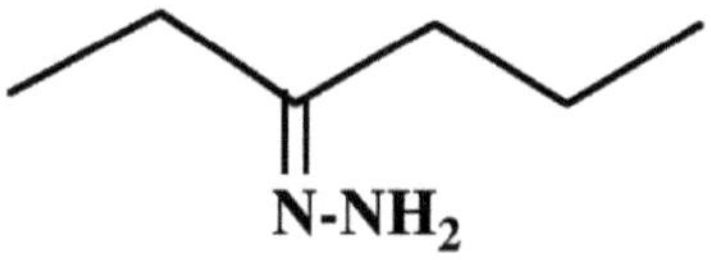

N-NH$_2$

E

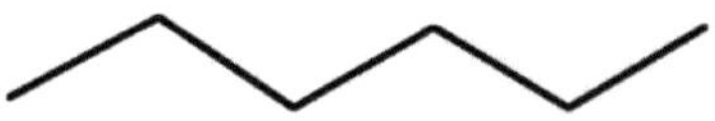

F

H_2/PtO_2

G

3.

(a) 1. $LiAlH_4$/ether, water work-up
2. $NaNO_2/H_2SO_4/H_2O$, 0°C
3. NaCN
4. Br_2/Fe

5. aq. acid, heat

(b) 1. aq. acid, heat
2. $SOCl_2$
3. $Li(CH_2CH_2CH_3)_2Cu$

(c) 1. $CH_3I/AlCl_3$
2. $CH_3CH_2C(O)Cl/AlCl_3$, water work-up

4.

1. treat the unknown tripeptide with phenylisothiocyanate

2. treat the modified tripeptide with trifluoroacetic acid and separate the resulting dipeptide from the phenylthiohydantoin derivative of the amino-terminal amino acid, comparing it to authentic phenylthiohydantoin produced from each of the three possible amino acids, phenylalanine, glycine, and alanine, using chromatography to establish the identity of the amino-terminal amino acid

3. treat the remaining dipeptide with phenylisothiocyanate

4. treat the modified dipeptide with trifluoroacetic acid and separate the resulting amino acid from the phenylthiohydantoin derivative of the amino-terminal amino acid, comparing it to authentic phenylthiohydantoin produced from each of the three possible amino acids, phenylalanine, glycine, and alanine, using chromatography to establish the identity of the amino-terminal amino acid

5. verify the identity of the carboxy-terminal amino acid by paper chromatographic comparison with authentic amino acid

5.

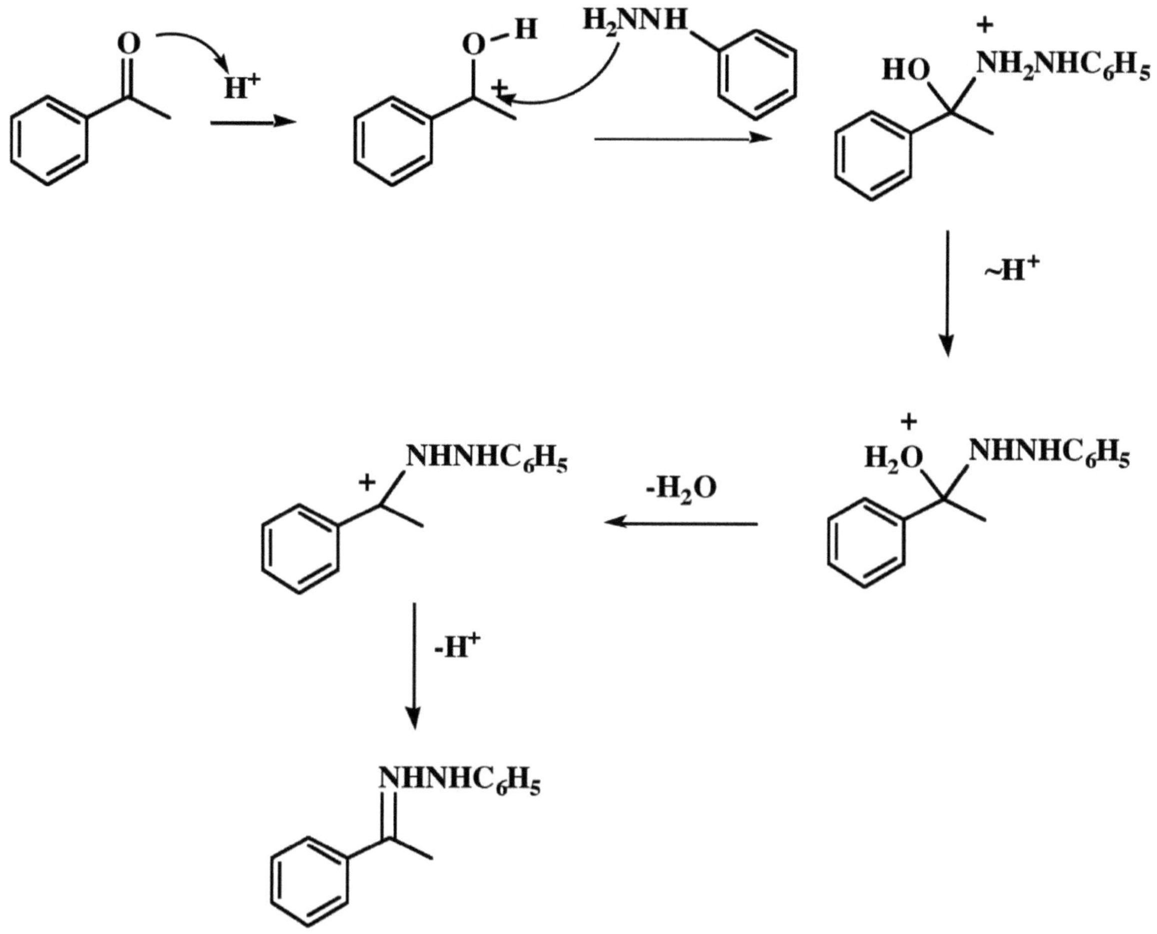

$$\text{Acetophenone} \xrightarrow{H^+} \text{protonated carbonyl} \xrightarrow{H_2NNHC_6H_5} \text{HO, } \overset{+}{N}H_2NHC_6H_5 \text{ adduct} \xrightarrow{\sim H^+} \overset{+}{H_2O}\text{, } NHNHC_6H_5 \text{ adduct} \xrightarrow{-H_2O} \text{carbocation, } NHNHC_6H_5 \xrightarrow{-H^+} \text{phenylhydrazone } (=NNHC_6H_5)$$

6.

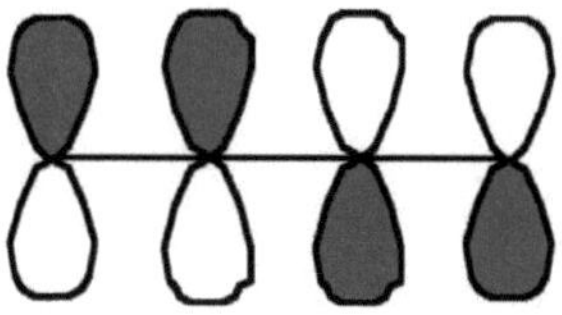

π_2 **HOMO of 1,3-butadiene**

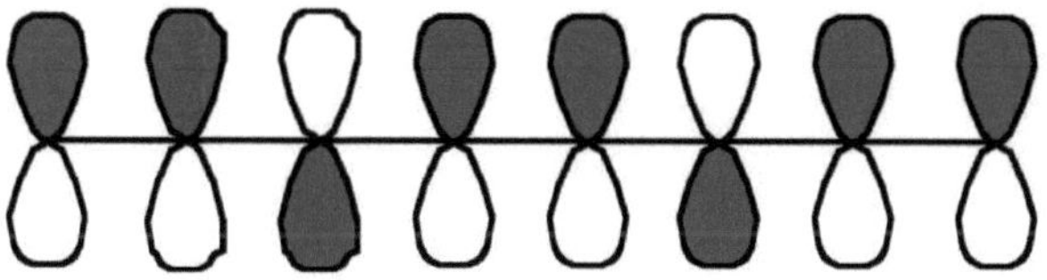

π_5 **LUMO of 1,3,5,7-octatetraene**

The reaction would not proceed feasibly under thermal stimulation, but rather under photochemical stimulation. It is more likely that an ordinary (thermal) Diels-Alder reaction would occur involving a portion of the conjugated π system of the 1,3,5,7-octatetraene.

7.

1. dissolve the mixture in an organic solvent such as hexane

2. wash the solution with aqueous sodium bicarbonate solution - The 4-ethylbenzoic acid will dissolve in it as the sodium salt leaving the remaining two compounds in the hexane layer. The 4-ethylbenzoic acid may be recovered from the aqueous solution by making it strongly acidic and extracting the free acid from it using hexane; drying the hexane extract and evaporating the solvent will provide the pure 4-ethylbenzoic acid.

3. wash the remaining hexane solution containing the two materials with dilute hydrochloric acid - The 4-ethylaniline will dissolve in it as the hydrochloride salt leaving the ethyl benzoate in the hexane. The ethyl benzoate may be recovered by drying the hexane solution and evaporating the solvent. The 4-ethylaniline may be recovered by making the aqueous solution strongly basic and extracting it with hexane. Drying the hexane layer and evaporating the solvent will provide the pure 4-ethylaniline.

8.

Measurement of the IR spectrum of each material in CCl_4 solution will provide the means for differentiation of the samples. The ethyl benzoate will exhibit no broad absorption through the range 3500-2800 cm^{-1} that is typical of carboxylic acids. The remaining two may be distinguished considering the region about 1200 cm^{-1}, within which the 4-ethoxybenzoic acid will exhibit very strong and sharp absorptions owing to the two ether linkages.

9.

(a)

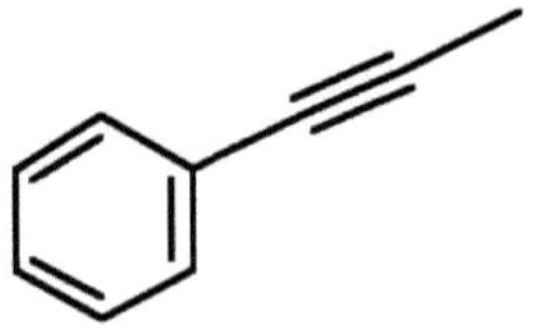

(b)

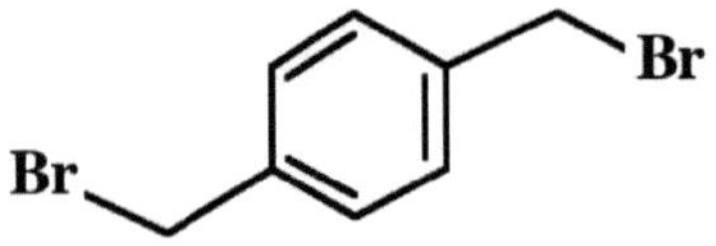

Br Br

1.

(a) the first compound exhibits a UV absorption at 278 nm while the second exhibits a UV absorption at 263 nm.

(b) the first compound exhibits a UV absorption at 245 nm while the second exhibits a UV absorption at 215 nm.

2.

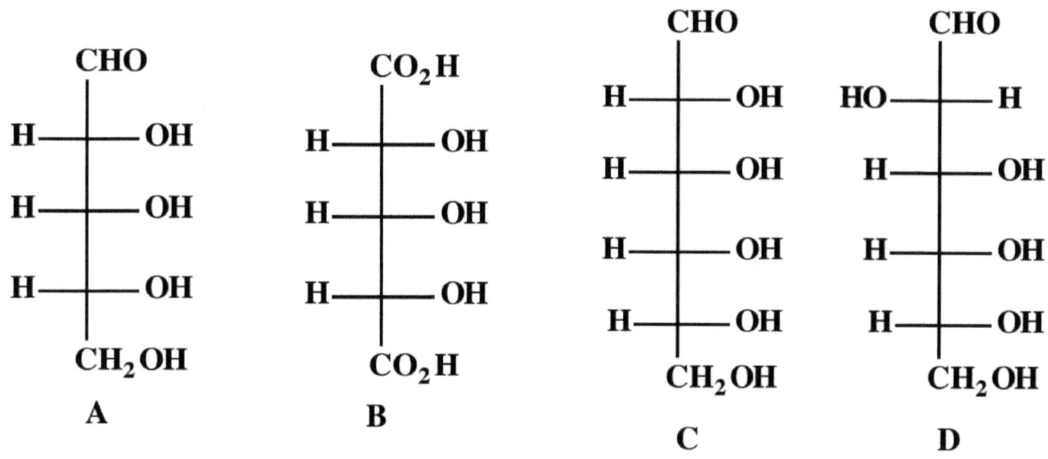

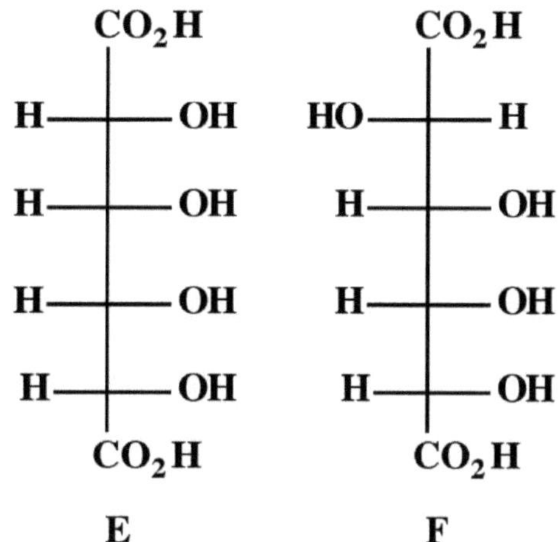

3.

(a) 1. $NaOCH_2CH_3$ to generate ethyl acetoacetate
2. $NaOCH_2CH_3$
3. CH_3I

(b) 1. $CH_3I/AlCl_3$
2. $KMnO_4/KOH/H_2O$ heat
3. H_2SO_4/HNO_3

(c) 1. $SOCl_2$
2. benzene, $AlCl_3$ - water work-up

(d) 1. triphenylphosphine
2. BuLi
3. cyclohexanone, heat

4.

We would anticipate cyclopentadiene to be a stronger acid than cyclopropene. On loss of a proton from cyclopentadiene, an aromatic anion is generated, whereas loss of a proton from cyclopropene will produce an anion in which electrons are unpaired and occupy antibonding molecular orbitals.

5.

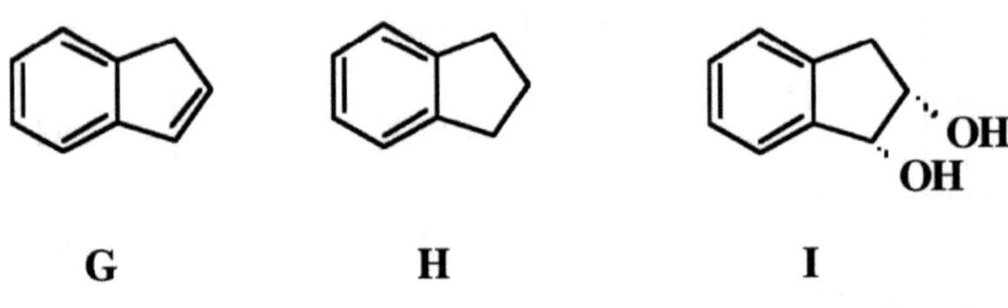

6.

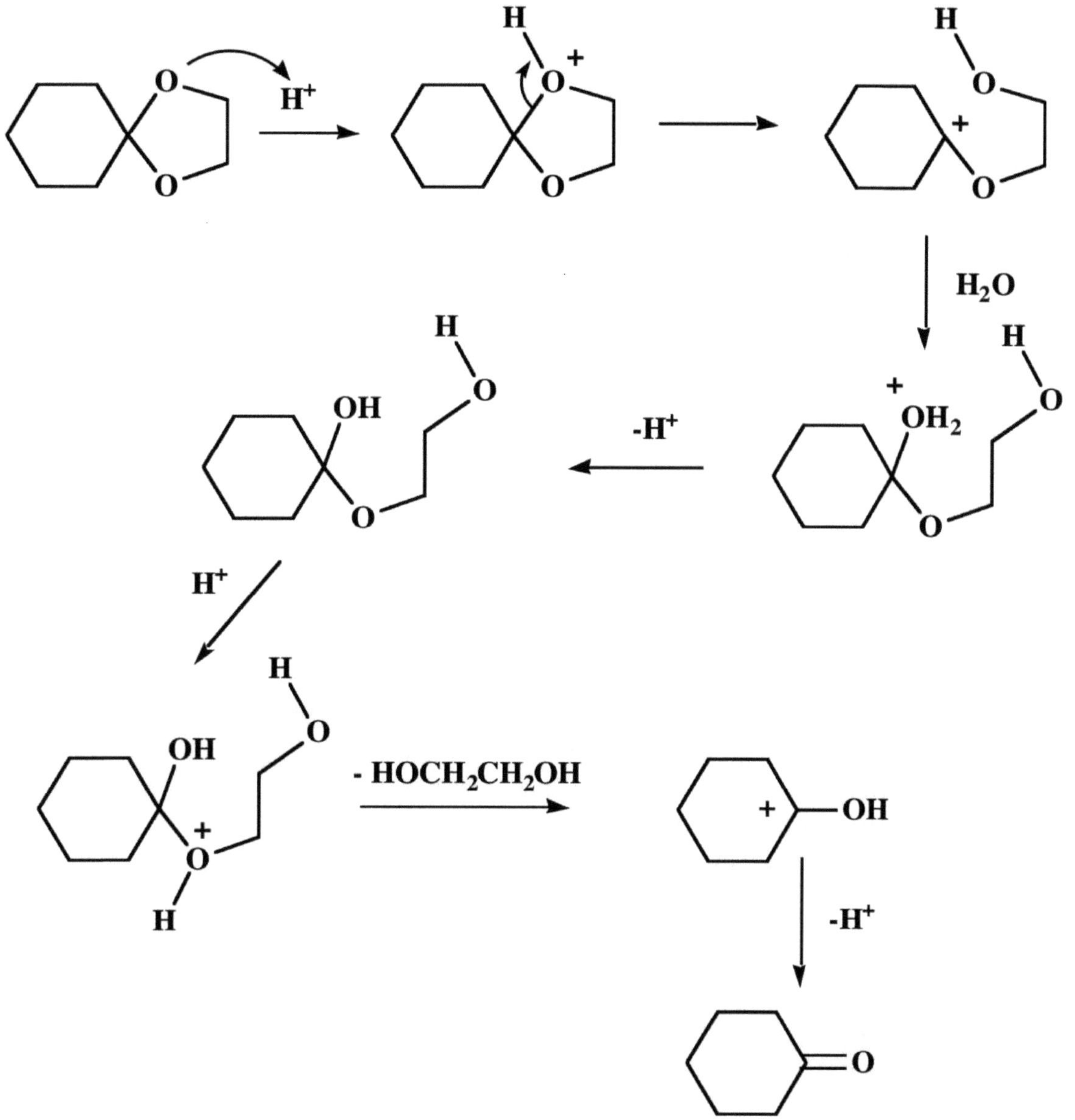

H⁺

H₂O

-H⁺

H⁺

- HOCH₂CH₂OH

-H⁺

7.

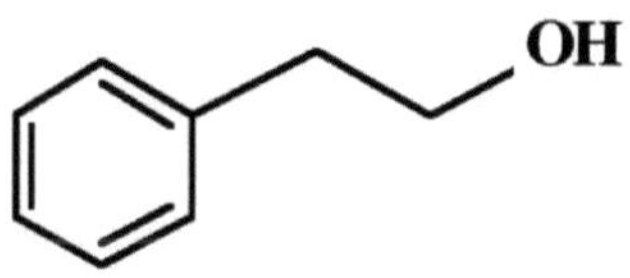

8.

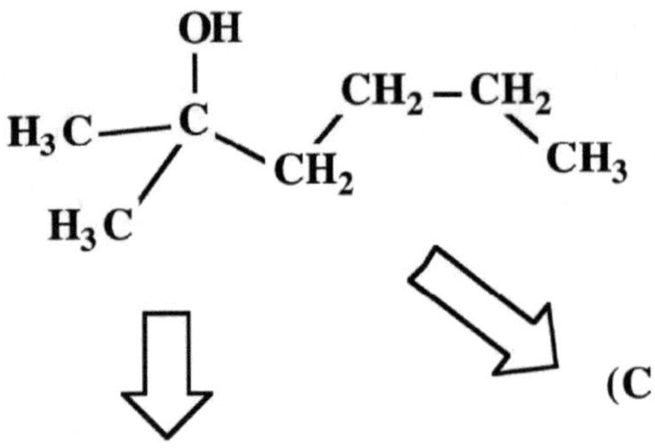

OH

$H_3C—C—CH_2—CH_2—CH_2—CH_3$ (with H_3C on C)

$(CH_3)_2C{=}O + CH_3CH_2CH_2CH_2MgBr$

$CH_3C(O)CH_2CH_2CH_2CH_3 + CH_3MgI$

9.

(a) 1. benzene is treated with Br_2/Fe
2. the bromobenzene is nitrated with H_2SO_4/HNO_3
3. the nitro group is reduced with Sn/HCl

(b) 1. Benzene is nitrated with H_2SO_4/HNO_3
2. the nitro group is reduced with Sn/HCl
3. the amino group is diazotized withn $NaNO_2/H_2SO_4$
4. $Cu_2(CN)_2$ is used to introduce the nitrile group
5. the nitrile group is reduced with $LiAlH_4$

(c) 1. benzene is nitrated with H_2SO_4/HNO_3
2. the product is brominated with Br_2/Fe
3.the nitro group is reduced with Sn/HCl

10.

(a) benzoic acid
(b) benzophenone
(c) $H_3CCH(NH_2)CN$
(d) $(CH_3CH_2)_2N\text{-}CH_2\text{-}CH_2\text{-}CHO$

ORGANIC CHEMISTRY II FINAL EXAMINATION (III)
ANSWERS

1. The 1,3,5-hexatriene would exhibit the smaller HOMO-LUMO gap. As the chain is extended to longer continuously conjugated linear systems, the HOMO-LUMO gap would continue to decrease, ultimately being located in the visible region of the electromagnetic spectrum.

2. a) the breakdown of sugars
   b) an elimination process leading to the thermodynamically less stable alkene product
   c) a carbohydrate of six carbon atoms bearing an aldehyde functionality
   d) a substituent on an aromatic ring that is withdrawing of electron density from the ring resulting in a less reactive ring toward electrophilic aromatic substitution reaction

3.

A

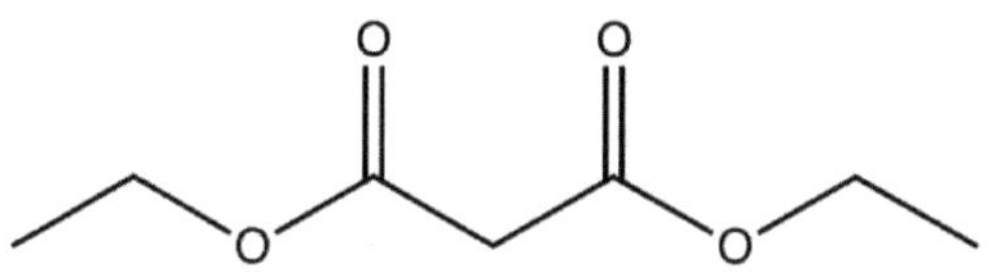

B

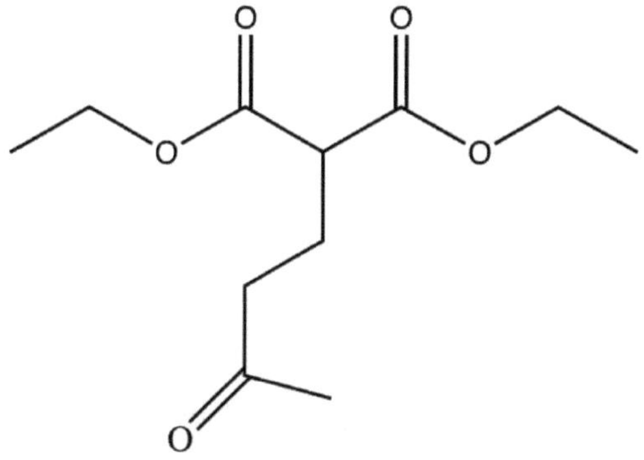

C

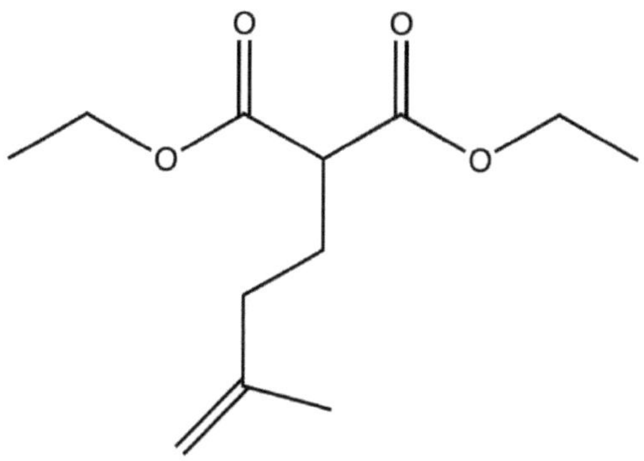

D

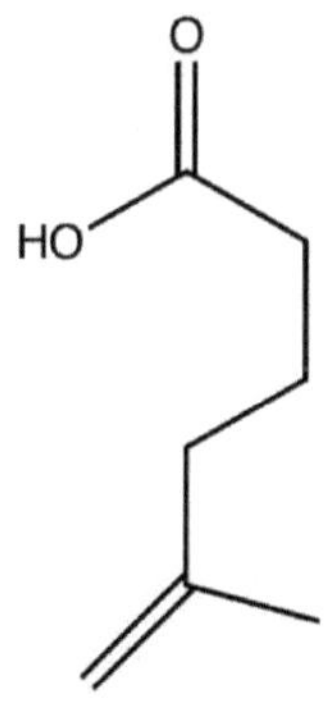

E

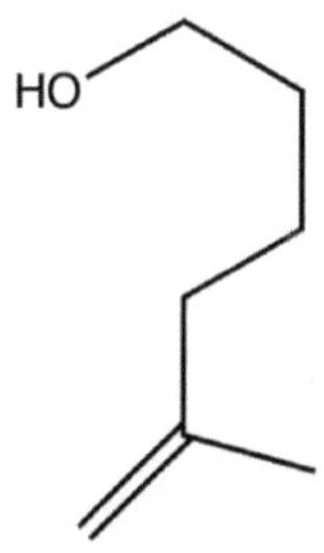

4. a)

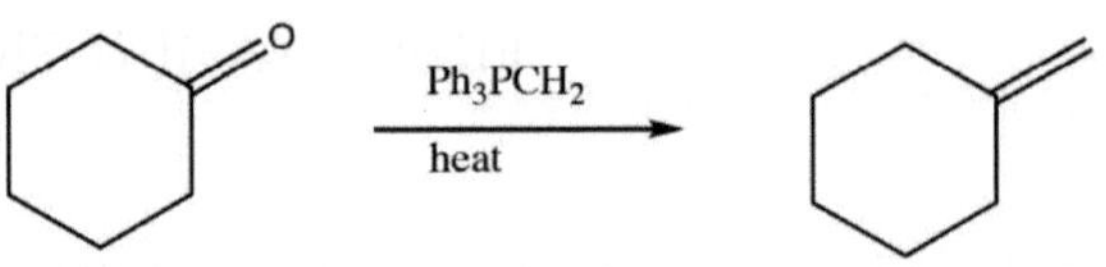

O

Ph_3PCH_2

heat

b)

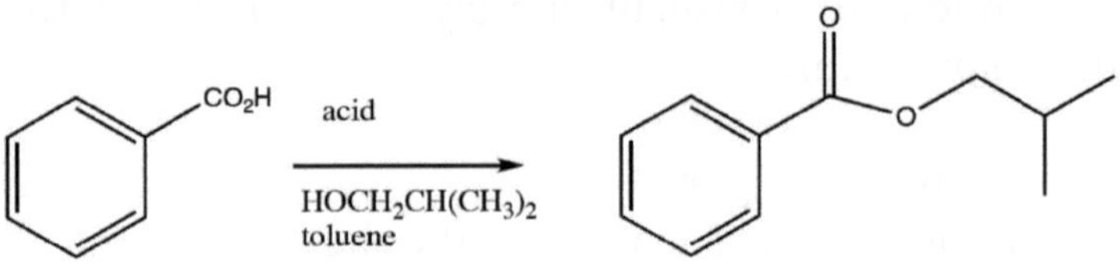

CO_2H

acid

$HOCH_2CH(CH_3)_2$
toluene

O

O

c)

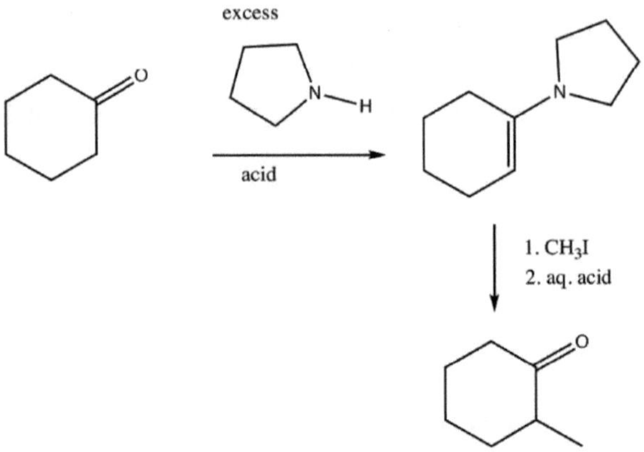

excess

O

N—H

acid

N

1. CH_3I
2. aq. acid

O

d)

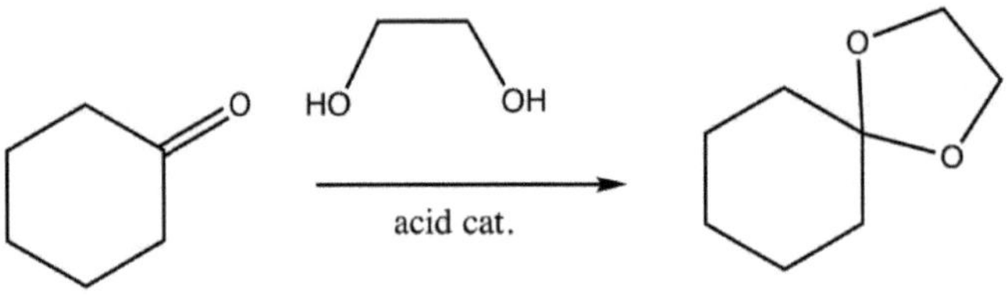

O

HO

OH

acid cat.

O

O

5. a) $NaIO_4$
b) 1. Benzoyl chloride; 2. Acetic anhydride; 3. Base; 4. Benzyl bromide; 5, aqueous acid
c) 1. NaCN; 2. Aqueous acid, heat; 3. Ethanol, acid cat., heat
d) 1. H_2SO_4, HNO_3; 2. $Cl_2/FeCl_3$; 3. Sn/HCl; 4. $NaNO_2/H_2SO_4$; 5. Cu_2Cl_2

6.

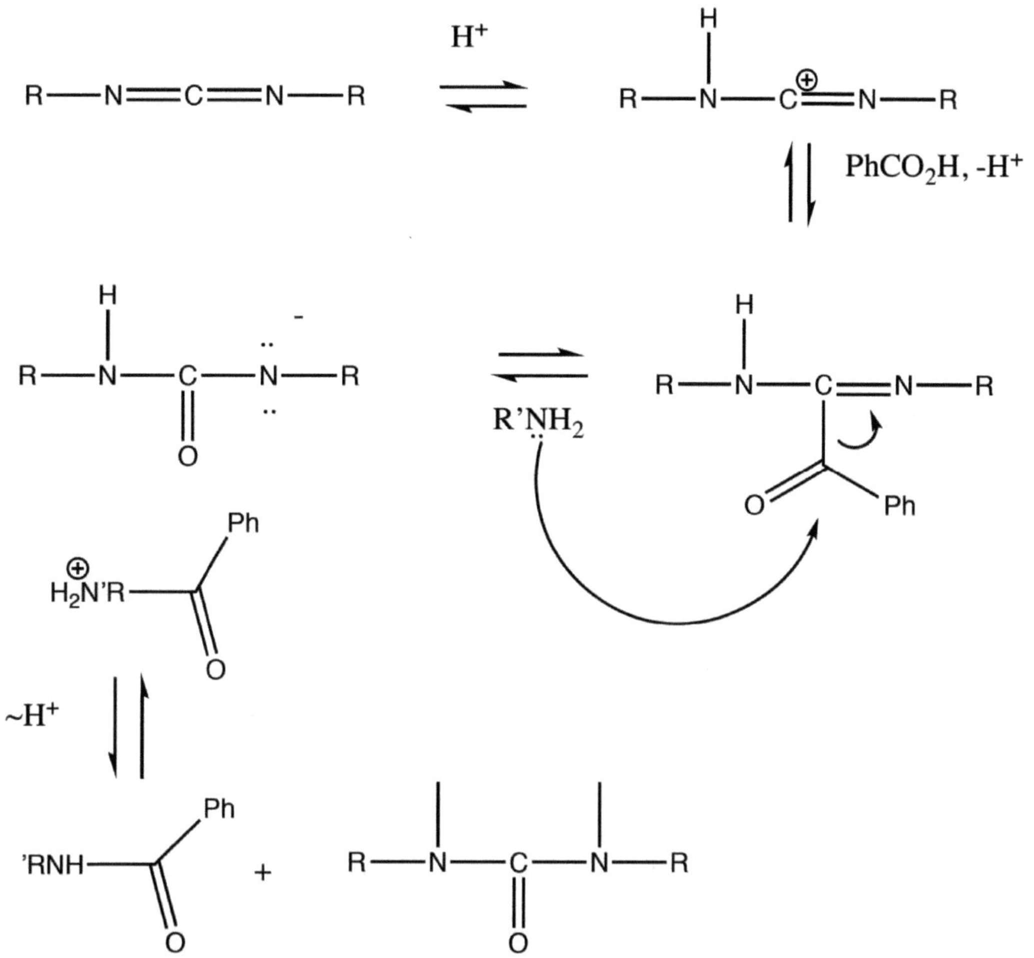

7. The reaction occurs by a nucleophilic *addition*of the methoxide ion to the carbon bearing the chloride, followed by the loss of a chloride ion from that same carbon atom. The reaction does *not* occur in the instance of the 2,4-dimethylchlorobenzene owing to the fact that the substituents on the ring are *not* electron withdrawing to allow the methoxide ion to add, whereas in the first instance, with 2,4-dinitrochlorobenzene, the substituents are highly electron withdrawing.

8. a)

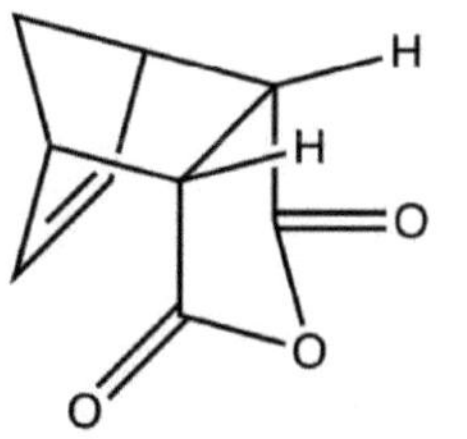

b)

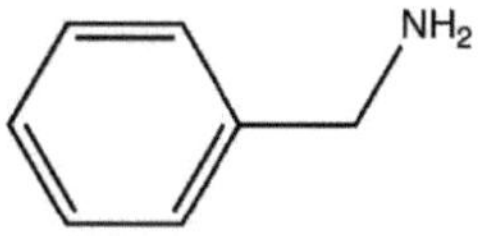

c)

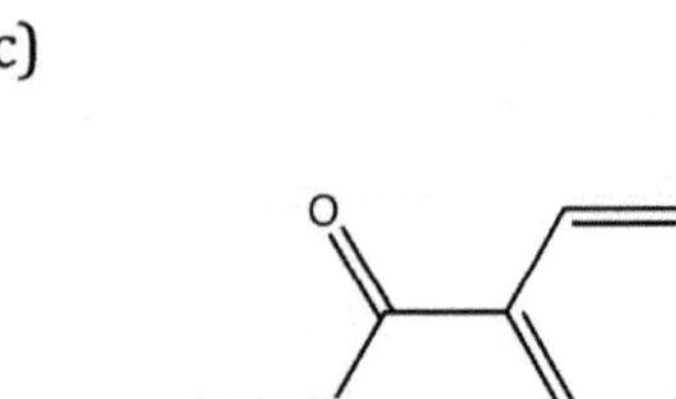

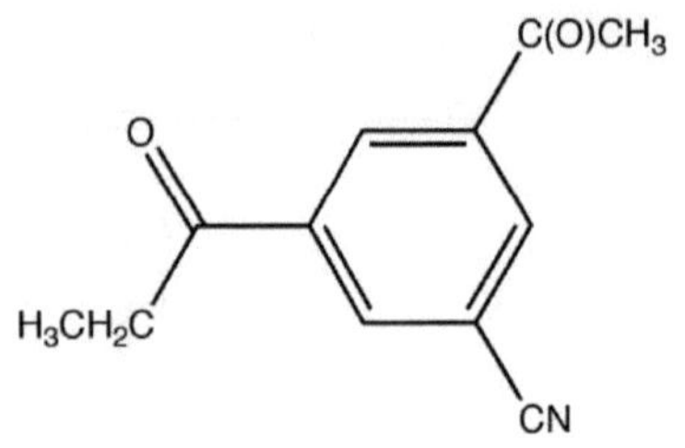

d)

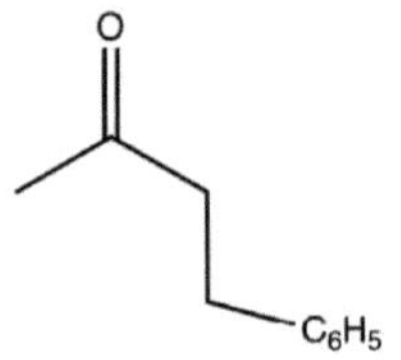

9. With the strongly electron withdrawing trichloromethyl group, the nitrile carbon is able to add the oxygen of the carboxylic acid to generate a type of mixed anhydride linkage as shown here:

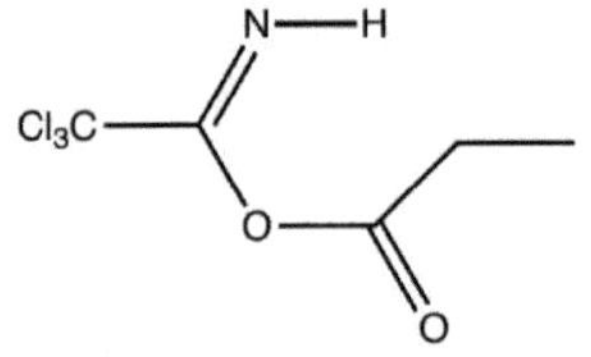

leaving the carboxyl carbon available for nucleophilic attack by the alcohol leading to the ester product along with the trichloroacetamide by-product.

10. Using the Wohl degradation:
1. H_2NOH, acid cat.
2. Acetic anhydride (excess)
3. Ammonia (excess)
4. Ag_2O

Printed by Libri Plureos GmbH in Hamburg,
Germany